# Diamonds
## in the Marsh

# Diamonds
# in the Marsh

*A Natural History of the Diamondback Terrapin*

BARBARA BRENNESSEL

With a Foreword by Bob Prescott

BRANDEIS UNIVERSITY PRESS
*Waltham, Massachusetts*

Brandeis University Press
© 2006 Brandeis University Press
Foreword © 2021 Robert Prescott
Preface © 2021 Barbara Brennessel
All rights reserved
Manufactured in the United States of America
Text design by Joyce C. Weston
Cover design by Cheryl Carrington
Typeset in Adobe Caslon, Formata, and Amazone

First Brandeis University Press edition 2021
Previously published by University Press of New England in 2006

For permission to reproduce any of the material in this book, contact
Brandeis University Press, 415 South Street, Waltham MA 02453,
or visit brandeisuniversitypress.com

The author and publisher gratefully acknowledge permission to reprint the
following: "The Turtle," copyright © 1930 by Ogden Nash, and reprinted by
permission of Curtis Brown, Ltd. "How the Partridge Got His Whistle,"
recounted in Canku Ota, 2003; http://www.turtletrack.org/Issues03/c00
/252003/CO01252003PartridgeWhistle.htm; used with permission.

*Library of Congress Cataloging-in-Publication Data*
Names: Brennessel, Barbara, author.
Title: Diamonds in the marsh : a natural history of the diamondback
    terrapin / Barbara Brennessel ; with a foreword by Bob Prescott.
Description: First Brandeis University Press edition. | Waltham,
    Massachusetts : Brandeis University Press, 2021. | Includes bibliographical
    references and index. | Summary: "Synthesizing all known research on this
    remarkable animal, 'Diamonds in the Marsh' is the first full-scale natural
    history of the diamondback terrapin. Focusing on the northern diamond-
    back, Barbara Brennessel examines its evolution, physiology, adaptations,
    behavior, growth patterns, life span, genetic diversity, land use, reproduc-
    tion, and early years"— Provided by publisher.
Identifiers: LCCN 2021019041 (print) | LCCN 2021019042 (ebook) | ISBN
    9781684580804 (paperback) | ISBN 9781684580811 (ebook)
Subjects: LCSH: Diamondback terrapin.
Classification: LCC QL666.C547 B74 2021 (print) | LCC QL666.C547 (ebook) |
    DDC 597.92/59—dc23
LC record available at https://lccn.loc.gov/2021019041
LC ebook record available at https://lccn.loc.gov/2021019042

5 4 3 2 1

# Contents

COLOR ILLUSTRATIONS FOLLOW PAGE 124

# Foreword

THIS BOOK IS ABOUT A TURTLE you will probably never see. Oh, they're out there if you know where to look, but they're really good at not being seen and they live in salt marshes. Even fisherfolks who make their living on the water rarely see them. But if you love turtles, biology, science, or a mystery, you're going to want to read this book. You'll learn all about how the natural history of this turtle was discovered.

You'll go metaphorically into the marshes and nesting areas with the biologists who have studied this species and have helped to peel back the cloak of mystery surrounding this shy salt marsh resident. The diamondback terrapin is an animal you'll want to know about. This charismatic turtle was pushed to the edge of extinction and slowly, very slowly, crawled its way to recovery.

As a sanctuary director, now semiretired, I had the wonderful opportunity to work on this rare, state-listed, handsome salt marsh turtle. The diamondback terrapin is unique in two ways: it's the only salt marsh turtle in the United States, and it reaches its northern limit in Wellfleet Harbor. I spent over 40 years at first just looking for terrapins, then buying and conserving land to protect their nesting areas, then researching and teaching about them, and finally developing ways to protect them.

It was in this capacity that I first met Barbara Brennessel over 20 years ago. She gave her daughter a gift to attend one of our weeklong terrapin field schools. Barbara joined her for that week of intensive classroom instruction, extensive field trips, and monitoring terrapins in a variety of locations at the sanctuary and around Wellfleet. At the end of that week our volunteers, our staff, our program, and myself would never be the same. While Barbara is an excellent geneticist and teacher, she is an even better field biologist and mentor. Switching from lab scientist to field biologist, she helped us all do a better job of learning about terrapins and ultimately protecting them.

One of the first field projects Barbara conducted was a genetic study of localized Massachusetts subpopulations of terrapins to see how they may be related, or not. Now that work is being included in a multistate study to look at how many species of terrapin there are from here to Texas. Barbara also set up studies that guided staff and volunteers through field trials for nest protectors that ultimately led us to nest-protection methods that don't cool the normal incubation temperature. That is important because soil temperature can affect the sex of the hatchlings. The book explains this phenomenon and many other discoveries that demonstrate what makes the diamondback terrapin so unique.

But this isn't a story just about terrapins in Wellfleet or on Cape Cod. The heart of the book explores the many research projects going on throughout the terrapin's range, from Massachusetts to Texas. It's also a story about young, aspiring biologists gaining experience in the field, investigating the so-called "lost years" of juvenile terrapins, practicing field techniques such as radio tracking, working in the lab learning how to head start baby terrapins, documenting differential rates of predation, and researching nest site selection and preferred substrates. There's a particularly interesting episode about the terrapins on the runways at John F. Kennedy International Airport in New York. Perhaps you've seen the photographs or read about them. The rest of this saga is a small part of the work going on in Jamaica Bay, an epicenter of terrapin research and conservation, conducted by professors and students at Hofstra University. Another epic conservation story highlights the research and actions being taken to prevent road kills. The Wetland Institute in Cape May, New Jersey, and the Georgia Sea Turtle Hospital on Jekyll Island, are inspirational examples that show what can be done when volunteers and scientists work together to mitigate human impacts on wildlife. These are only a few examples; there's so much more that makes this book essential reading for any turtle enthusiast: sex life, diet, brumation, nest site selection, and predation.

Why is the protection and preservation of the diamondback terrapin so consequential? Because terrapins are a significant species, situated at the top of their food web. They feed on predatory snails and crabs, crabs that feed on clams, quahogs, and the famous Wellfleet oyster. They are a critical component of healthy ecosystems. They are great neighbors. But they face many manmade challenges, including drowning in crab traps, loss of nesting habitat through natural succession and coastal development, illegal harvest for food and the pet trade, and just plain ignorance on our part.

We all need to be stewards and champions of this turtle. Learning about its natural history, reading about the researchers, young and old, studying an ancient species, trying to bring it back from the edge of extinction in so many places, will inspire and motivate you to become a more forceful advocate for terrapins. You can help by joining a volunteer group in your state or town or at your local nature center. Sign up to be a member of the Diamondback Terrapin Working Group to learn more, to get involved and become a knowledgeable defender. Rebuilding their populations so they can resume their role in the ecosystem is the ultimate goal. We're getting there. The diamondback terrapin has a much brighter future now than when I started my work, and the story told in this book explains why.

Bob Prescott, Sanctuary Director Emeritus
Mass Audubon's Wellfleet Bay Wildlife Sanctuary
February 22, 2021

# Preface

SINCE THE INITIAL PUBLICATION of *Diamonds in the Marsh: A Natural History of the Diamondback Terrapin*, I have observed slow, steady progress in the realm of diamondback terrapin research, conservation, and education. This progress is facilitated by the participation of an increasing number of conservation-minded institutions, organizations, agencies, and teachers, another generation of research students and interns, and an upsurge in the number of naturalists and volunteers who are captivated by this salt marsh turtle.

Our knowledge of terrapin natural history and ecology has modestly expanded: we now have further documentation of a terrapin colony in Bermuda, inhabiting brackish ponds, some in an area near a golf course. This population resides 2,873 km (1,875 miles) away from the continental United States. There are conflicting reports about how these terrapins managed to get there and how long they have been in residence. Terrapins were transported in large numbers for food, so perhaps the Bermuda terrapins are descendants of escapees, washashores from a ship that visited the island. Perhaps they were brought there intentionally as pets or for food. Various studies using fossil, carbon dating, environmental, and genetic data suggest that they may have arrived as early as 3,000–4,000 years ago, or as late as the 1600s, perhaps by riding the Gulf Stream from the Carolinas. Thus, the jury is still out about the origin of Bermuda terrapins.

Genetic analyses are more sophisticated today. Molecular techniques are further refining subspecies delineations. Using a spectrum of genetic markers, investigators are documenting multiple paternity in all terrapin populations in which it has been studied. Additional terrapin clusters are known, and population studies using high-tech methodologies are providing more accurate assessments of population sizes and geographical distribution.

As a species, and also for turtles in general, terrapins exhibit a number of biological adaptations such as temperature-dependent sex determination, hibernation (also known as brumation), and estivation. We now have additional insights into the physiological and biochemical basis for these and other unique terrapin life-history traits (as summarized in W. M. Roosenburg and V. S. Kennedy, eds., *Ecology and Conservation of the Diamond-*

*Backed Terrapin*, 2018, Johns Hopkins University Press). But many questions still remain.

Progress has also occurred on the conservation front. Terrapins have been placed on the International Union for the Conservation of Natural Resources (IUCN) list as globally near threatened, and in 2013, they were added to the CITES (Convention of International Trade and Endangered Species) list due to concerns about international trade. States that once permitted limited harvest—Connecticut, New York, New Jersey, and Virginia—now have full bans. As part of the development of state and regional conservation plans, there are increasing efforts to prevent depredation of nests, to avert road mortality, to inhibit bycatch in commercial and recreational fisheries, and to restore terrapin habitats.

The Diamondback Terrapin Working Group (DTWG), a consortium of researchers, conservationists, and educators, has taken the lead in advocating for the prevention of terrapin bycatch and subsequent mortality in the blue crab fishery. In 2020, the DTWG membership voted to adopt a "Position Statement on the Negative Effects of Blue Crab Traps/Pots on Diamondback Terrapin Populations and the Use of Bycatch Reduction Devices as a Practical, Inexpensive Solution."

The story of one terrapin, Bayley, is an example of the far-ranging communication and cooperation among terrapin advocates, which has been fostered by the DTWG. In October 2019, I had just returned home to Wellfleet following a road trip to Wilmington, North Carolina, to attend the Eighth Symposium on the Ecology, Conservation and Status of the Diamondback Terrain. I checked my e-mail and there was a message from Dr. Charles Innis, veterinarian at the New England Aquarium. Dr. Innis had been contacted by a veterinarian from Maine who had just x-rayed a female diamondback terrapin. The terrapin was purchased at a reptile expo and the new owner wanted to know if the turtle bore eggs. Instead of potential "bundles of joy," the x-ray revealed a microchip, a Passive Integrated Transponder (PIT). With a special reading device, the Maine veterinarian was able to detect the unique PIT tag number of her patient. Dr. Innis asked me if I would contact fellow terrapin researchers to try to find the origin of the turtle. I first touched base with local colleagues and eliminated Massachusetts terrapins from the likely suspects. I then put out a notice on the DTWG e-mail list and got a hit. The terrapin from Maine had been part of a population study in Barnegat Bay, New Jersey, conducted by researchers from Drexel University. It turned out that she was illegally collected, and had traveled through several states before she turned up at the reptile expo in Maine.

The terrapin's owner wanted to do the right thing, and thus the return of the terrapin to New Jersey was eventually orchestrated by several agencies, including New Jersey Fish and Wildlife, the Conserve Wildlife Foundation of New Jersey, and the Turtle Survival Alliance. The turtle was received by John Wnek, program terrapin coordinator and supervisor of science and research at the Marine Academy of Technology and Environmental Science. Wnek reports that Bayley was subject to a reptile physical exam consisting of a complete health assessment and tests for pathogens. She was quarantined at the Nature Center at Island Beach State Park while awaiting potential release into her native marsh. It took a strong network of collaborators to rescue this single terrapin.

Many challenges remain in the area of terrapin conservation. Subsequent to the frenzy in coastal development that occurred in the twentieth century, the loss of terrapin habitat (i.e., coastal areas and salt marshes) is slowing down dramatically, as there is very little pristine coastline remaining. However, residual habitat is being impacted by the effects of climate change. We are already witnessing sea level rise and an increase in the number and severity of storms that whip up waves and wind. Our marshes are in danger of disappearing and suitable terrapin nesting areas have become inundated or have shriveled in size. However, I am encouraged to see progress in the area of habitat stabilization and restoration. There are many projects, large and small, that contribute to terrapin conservation. Creation of terrapin nesting areas, also known as turtle gardens, has become increasingly popular to mitigate for untoward effects of human activities and climate change. Turtle gardens can also prevent road mortality when they are strategically placed near terrapin road-crossing areas. They entice the females to nest before they have a chance to wander onto the roads.

One of the larger-scale habitat conservation undertakings is salt marsh restoration. This type of project provides habitat not only for terrapins but also for many other salt marsh denizens. The Herring River Restoration in Wellfleet is a case in point. Briefly mentioned in the first edition of this book, this project has been in the works for almost twenty years. I have monitored the terrapins in the Herring River for ten years and witnessed the decline of suitable nesting habitat and degradation of water quality over that period of time. But we are inching closer and closer to witnessing the return of tidal flow, the improvement in water quality, and eventual salt marsh restoration. As an example of the complicated nature of these projects, it has taken dozens of scientists and over a decade of work, including monitoring, environmental studies, engineering, design, permitting, cooperation between local, state, and federal agencies, a memorandum of understanding between two towns (Wellfleet and Truro) and the National Park Service, a dedicated nonprofit agency

(Friends of Herring River), dozens of organizations working to support the project, and settlement of potential conflicts of interest with property owners, shellfishers, wildlife enthusiasts, recreational users of the area, and others, not to mention the $50 million price tag.

Although this book is about a single species, we know that this species must be studied and protected in a larger ecological context. The attention given to terrapins has not abated; in fact, it has increased. My hope is that this book will continue to encourage interest in terrapins and foster the protection of this species and the habitats where they reside.

# Acknowledgments

Don Lewis provided the inspiration for this project. I suggested to him that he should write about diamondback terrapins using his Internet stories as a starting point. When Don's myriad activities kept him busy with other endeavors, I decided to plunge into the project. Bob Prescott provided encouragement as I described my idea for a book on terrapins while he performed a necropsy of a two-year old loggerhead turtle that was found dead near Martha's Vineyard. The idea for the book on terrapins became solidified as we examined the loggerhead, inside and out, for any signs that would explain its demise.

I visited many diamondback terrapin habitats along the way and met many researchers who gave generously of their time and expertise: Roger Wood, Christina Watters, Roz Herlands, Brian Mealey, Greta Parks, Marguerite Whilden, Matt Draud, Barbara Bauer, Charlotte Sornborger, Mike Ryer, Sue Nourse, Amanda Widrig, Russ Burke, and Peter Auger. Kristin Hart and Susanne Hauswaldt shared DNA primers and their genetic expertise. I spoke to many Department of Natural Resources (DNR) and Fisheries and Wildlife biologists to discuss terrapin conservation and regulations.

William E. Davis provided the drawings throughout the book. I owe him many thanks for volunteering to illustrate the manuscript, for his simple, clear, and expressive depictions, and for his trips to the Harvard Museum of Comparative Zoology to study the archived terrapin specimens, most of which were victims of crab pots.

Bob Prescott, Russ Burke, and an anonymous reviewer provided many important and useful suggestions. They helped to fill in many gaps in my self-acknowledged deficiencies as a saltmarsh ecologist and herpetologist. I am solely responsible for any errors that remain. Nicholas Picariello, Jilann Picariello, Scott Shumway, Shelly Leibowitz, and John Kricher read drafts of portions of the book and provided wonderful suggestions. Wheaton College librarians T. J. Sondermann and Martha Mitchell helped to track down obscure reports, theses, and anecdotes.

Many Wheaton students participated in field and laboratory projects described in this book: Cate Hunt, Rob Monteiro, Cait Stewart-Swift, Nick Warren, Joe Chadwick, Kate Leuschner, and Lindsey Shorey. In a pilot headstarting program at Wheaton College, students assumed responsibility for feeding and care of hatchlings: Cate Hunt, Lindsey Jordon, Ashley Jennings, Orissa Moulton, Amy Brown, and Kara Marquis. I am particularly indebted to Joe Chadwick for training students, setting up feeding schedules, proposing improvements, and taking a key role in the project. Liz Jacques and Diana Page conducted molecular studies during their senior year at Wheaton. My daughter Adriana Picariello assisted with fieldwork for two years. Meghan Walsh was a terrapin intern while she studied biology at Skidmore College. Many Wellfleet Bay Wildlife Sanctuary volunteers walked miles and miles in nesting areas to look for signs of nests and hatchlings and to check protected nests.

Erin Post and Kathleen Morgan were most helpful with suggestions and assistance with terrapin husbandry. Kathy Rogers provided expert assistance in assembling the manuscript. Funding for terrapin field and lab projects and for travel and production of this book was provided by a Wheaton College Faculty Research Award and Goldberg Chair stipend. Wheaton Foundation Awards and Mars Internships were awarded to many of the students. Nest monitoring and tracking studies of yearling terrapins in 2003 were funded by the Sounds Conservancy Grant Program.

I give special thanks to my editor, Phyllis Deutsch, for her newly acquired interest in diamondback terrapins, her faith in the value of a single-species book that could also unfold a story about humans and their environment, and her enthusiasm for this project.

FAST FORWARD TO 2021. Another generation of interns and countless volunteers have participated in some of the work herein described. I was not able to find print copies of this book to give them as background for their terrapin involvement. I thought I would never see a printed version of this book again. But then, quite unexpectedly, I found my publishing guardian angel, Sue Ramin, director of Brandeis University Press. Sue is another proponent of the terrapin story who shared my excitement about reviving this print edition of *Diamonds in the Marsh*. To Sue, I will be forever grateful.

Again, I thank Bob Prescott, not only for his foreword to this printing, but also for his continued leadership in terrapin research, conservation, and education. Bob's recent retirement as director of Wellfleet Bay Wildlife Sanctuary has afforded him even more time to guide and direct important initiatives for our Cape Cod turtles (our local terrapins and box turtles as well as cold-stunned sea turtles), and he has truly plunged right in.

# Introduction

THE CALL TO ACTION came via e-mail. After the vernal equinox announced the arrival of spring, veteran terrapin researcher Don Lewis, the "Turtleman of Wellfleet," had monitored daily air and water temperatures. During the preceding week, a significant warming trend was observed. It was time to see if the diamondback terrapins had begun to parade in Blackfish Creek. Twice a month, an hour before each spring tide, the creek becomes passable on foot. Armed with a landing net and protected from the chilly April waters with a pair of thermal waders, I joined Don to trek across the muddy tidal flats into the main channel of the creek. The wind from the northeast whipped up the shallows, limiting our visibility to only a few inches. Our fingers were gripped around the handles of our nets, frozen into place as we waited patiently for the dim silhouettes of drifting turtles.

Soon, we could see their heads: larger heads for the females, smaller ones for the males and juveniles. The terrapins were being flushed from the smaller, innermost creeks into the main channel of Blackfish Creek by the ebbing tide. They bobbed along with the current, occasionally periscoping their heads above the water for a breath of air and maybe a view of their destination ahead. These terrapins had left their winter homes, crypts in the muddy creek bottoms, and had begun to make their twice daily tidal journey downstream, with a return trip upstream, in the creeks of Wellfleet harbor.

Dip netting in Blackfish Creek is an inefficient operation at best and a complete folly when the wind is howling and the air and water temperature are borderline freezing. Standing in water, knee to thigh deep, one can only hope to see a familiar form swimming within netting range as the 8 to 10 foot drop in tide whisks terrapins away from their shallow upstream locations. Occasionally, we are rewarded by a "thunk," as a terrapin has crashed into one of us, or, even better, has barreled into our net. Sometimes, we actually see and catch one. That April day, weather conditions had turned gloomy. We were cold, wet, and tired, but had nevertheless been rewarded by witnessing the terrapin parade. We knew the turtles had awakened from their winter slumber and were active again. As we prepared to call it quits, I saw a familiar form scooting by my ankles. I swooped down with the net and felt the extra heft as I lifted it from the water. I optimistically peered down and saw her. She was a large turtle, a familiar one that we had captured two years earlier. Female number 1007 heralded the arrival of another terrapin season.

Getting wet and muddy in the name of science represents a genuine departure for me. Trained as a biochemist and molecular biologist in the pre-cloning era, I had spent most of my career in a white coat at a laboratory bench or in front of a classroom filled with undergraduates. My research was focused on topics with potential applications to human health: the mechanism of action of peptide hormones and the development of fat cells. But the more time I spent outdoors in New England, the more I became concerned about the environment in which my husband and I were raising our children. What good would it be to contribute to our knowledge of human health if we were destined to live in an unhealthy environment?

Many of my friends who had worked in the corporate world had experienced mid-career changes. I asked myself, "Why couldn't a biochemist take up the challenge of working on a project that might also help to preserve the environment?" As a summer resident of Wellfleet, a small Cape Cod town, I was interested in preserving the nature of the Outer Cape, an endeavor that was also important to the late President John F. Kennedy when he created the Cape Cod National Seashore. As a part-time Wellfleetian, I shared my summers with a number of creatures that are part of the fragile landscape, and I endeavored to learn more about them. This task was made more urgent after my introduction to *Malaclemys terrapin*, the northern diamondback terrapin.

I had become familiar with our local population of Eastern box turtles, painted turtles, and snapping turtles. Box turtles can be found in their characteristic cavelike forms in the dense pine-needle underbrush; painted and snapping turtles inhabit freshwater ponds and creeks and are plentiful in Eel Creek, on a border of our property. All three species nested in our sandy driveway and under our clothesline. We sometimes found a straggler trapped in one of our window wells. My children spent their summers observing our reptilian neighbors, and occasionally we would come upon hatchlings that had overwintered in their nests. We were even fortunate to get a glimpse of sea turtles during fishing expeditions in Wellfleet Harbor and Cape Cod Bay. But it wasn't until I attended a field research course with my oldest daughter at the Wellfleet Bay Wildlife Sanctuary (WBWS), a division of the Massachusetts Audubon Society, that I became aware that I was living in the midst of yet another type of turtle, the diamondback terrapin. After one summer of participating in diamondback terrapin research and conservation efforts, I was hooked! This was a turtle that could use more friends. And so, my mid-career change had begun.

I easily recruited undergraduate student interns from Wheaton College who were excited about the prospect of summer field studies on Cape Cod. We partnered with Wellfleet Bay Wildlife Sanctuary and contributed to its long-term population study of diamondback terrapins in Wellfleet Harbor. The Sanctuary helped to provide housing for interns during the busy summer tourist season. When housing was impossible to find, interns bunked at our house and sometimes slept in tents in our yard. We kayaked, forded creeks, got stuck on mudflats and in ooze that sucked the boots from our feet, and walked many miles of marsh and dirt roads.

We captured terrapins from creeks and coves, followed nesting females and protected their freshly laid eggs, took blood samples for genetic analysis, and participated in WBWS education and outreach activities. Our trips back to Wellfleet in September and October were rewarded by the sight of baby turtles. We came full circle to witness the entire annual activity cycle of Wellfleet terrapins. While they hibernated, we went back to the laboratory and isolated DNA from blood samples and performed genetic analyses.

The finding of terrapin eggs one late October after a fierce nor'easter provided additional opportunities for study. Wind and waves had eroded the home of a future generation of terrapins and exposed the eggs to the elements and to predators. Since it was so late in the season, the viability of the embryos was questionable. The eggs were placed in a bucket of moist sand, and within days, tiny terrapins emerged. Normally, the hatchlings would be released, but these neonates had already survived one close brush with death, and the winter weather was upon us. Freezing temperatures could easily cause their demise. With proper permits from the Massachusetts Department of Fisheries and Wildlife in hand, we brought the hatchlings to the laboratory at Wheaton. Thanks to Dr. Peter Augur, who was also raising terrapins in laboratories at Barnstable High School and Boston College, we received an accelerated lesson on hatchling husbandry. We successfully stewarded the hatchlings through the winter with a warm home, plenty of food, and devoted undergraduate caretakers. This headstarting experience gave us the chance to observe hatchling growth and behaviors and to track the young terrapins when they were released into their natal marshes during the following spring.

As a growing number of interns and volunteers requested background reading information for their terrapin fieldwork and as I endeavored to learn more about these shy turtles, I spent quite a bit of time digging for information. I realized that much of the recorded natural history of diamondback terrapins, including historic records and more recent scientific studies, was scattered in scientific journals representing a variety of fields of study and in various reports written for state, local, and private, nonprofit organizations and agencies. I recognized how useful it would be for naturalists, researchers, and environmentalists to have a summary of the information collected in one volume. I decided that this book would be my contribution to the conservation of this elusive species and its rapidly eroding habitat. In addition to describing the natural history and ecology of the diamondback terrapin, I endeavored to trace the intersection of local terrapin populations with the history of the settlement and development of coastal areas.

The tale of the diamondback terrapin cannot be complete without an account of exploitation of this turtle by humans and the challenges to its recovery presented by new and continuing pressures. Conservation assessments point to the current need to develop a proactive stance to protect this species from a declining population trajectory. Perhaps it is a tale with a potential for a happy ending if current threats to the species can be mitigated and new threats can be prevented.

# Diamonds
## in the Marsh

# Chapter 1

# A Decidedly Unique Creature

"TERRAPIN" is a term from the Native American language to describe edible turtles that live in fresh or brackish waters. One such species is *Malaclemys terrapin*, the diamondback terrapin, the only U.S. turtle known to inhabit estuaries, coastal rivers, and mangrove swamps. The diamondback terrapin is unique. It tolerates fresh water, salt water, and everything in between. Unlike other turtle species that are exclusively freshwater or marine, the diamondback terrapin actually prefers an environment with intermediate salinity. This medium-sized turtle is confined to the Atlantic and Gulf coasts of the continental United States and can be found from Cape Cod, Massachusetts, the northernmost range of the species, to Corpus Christi, Texas. Seven subspecies have been identified.

The diamondback terrapin gets its name from the raised concentric rings formed on the subsections of the shell or carapace. The resulting grooves give the shell a sculpted appearance and are reminiscent of the facets on a cut diamond. The origin of the genus name, *Malaclemys*, is a bit obscure. An initial attempt at etymology would suggest that this is a genus of "bad" turtles. However, the genus has had several names, one of which was *Malacoclemys*. This was apparently shortened to *Malaclemys* at the end of the nineteenth century. The Greek "malakos" means "soft," and "klemmys" is a Greek word for turtle. In one of the initial descriptions of the genus by Gray in 1844, it was noted that the turtles had soft, spongy heads. However, "mala" apparently refers neither to their temperament nor their spongy heads. Ernst and Bury (1982) posited that "soft" may relate to their soft-bodied molluscan prey. Diamondback terrapins are therefore "mollusk-eating turtles" rather than "evil" turtles.

Turtles are unusual creatures. There are many characteristics of turtles that cause them to stand out from most animals and even from other reptiles.

3

Undoubtedly, the turtle shell is the most noticeable and most unique feature of these animals. Another important feature, less noticeable and less unique, is the fact that most turtles, including diamondback terrapins, are ectothermic animals. Similar to other reptiles, they do not have a physiological mechanism to maintain a constant body temperature and must rely heavily on their surroundings and behavioral adjustments to provide suitable temperatures to carry out physiological and cellular functions. Another interesting biological idiosyncrasy that sets turtles apart from many other familiar animals is a phenomenon called temperature-dependent sex determination (TSD). Most turtles lack sex-determining chromosomes such as the X and Y chromosomes of many species. A turtle is destined to become a male or a female depending on the temperature at which it develops inside its egg. It is interesting to learn about how these reptiles appeared and how they have managed to persist, despite, or perhaps because of, their peculiar characteristics.

## Evolution and Classification

Shelled reptiles are known by several names. Historically, they have been called tortoises if they are land-dwelling, turtles if they are aquatic, or terrapins if they belong to certain edible species. They are collectively known by the proper scientific name, "Chelonians," derived from another Greek word describing turtles. The Greeks also used the word to designate a battle formation in which soldiers marched against the enemy with interlocking shields. "Testudo" was also used to describe this type of formation. This "engine of war" would indeed physically and functionally resemble the upper shell or carapace of a turtle. Despite the diversity within the Chelonians, we commonly use the term "turtle" as an all-inclusive identifier. How did they originate, these strange creatures that carry their homes on their backs? We can trace the evolution of the turtle back about 200 million years, during the Triassic period, when *Proganochelys* made its appearance on the earth. There is a dearth of evidence to tell us the complete story of the transition of some of the early reptiles into turtles, but the oldest turtle fossils were found in modern-day Germany, Greenland, and Thailand. Evolutionary biologists believe that turtles evolved from small reptiles and may have retained many of the features present in their ancient predecessors. There is some speculation among scientists that turtles arose from the ooze: marshy areas or swampland between terra firma and fresh bodies of water. Based on several criteria, Reippel and Reisz (1999) postulated an aquatic origin for turtles. The type of respiration and locomotion exhibited by turtles could probably be achieved more easily in the water, where buoyancy could support the extra weight of the shell and limb muscles could assist with breathing. The presence of a bony plastron, the lower shell, makes sense for an aquatic organism that would need protection on its ventral, or lower, body surface. It could be argued that there is less exposure of the ventral surface of four-legged land-dwelling animals to possible predators.

The development of a boxlike shell undoubtedly gave turtles protection from sharp-toothed and/or strong-jawed predators. No matter how it occurred, the development of a shell was so successful that the basic components of the turtle body plan have changed very little over time. A major difference between the first turtle, *Proganochelys*, which was about 0.6 meters (2 feet) in length (Alderton, 1988), and the modern Chelonian is the modification of a jaw that originally contained teeth into a horny, beaklike jaw. During the Jurasssic period, 210 to 144 million years ago, some turtles moved away from the swamps or their freshwater homes, into the oceans and onto the land. At the end of the Jurassic, turtles had developed a flexible neck and could retract their heads into their shells. During the Cretaceous period, 144 to 65 million years ago, a huge turtle named *Archelon*, 3.6 meters (12 feet) long, roamed the oceans. During the mass extinction at the end of the Cretaceous, *Archelon* went the way of the dinosaurs, but some of his smaller turtle relatives survived the catastrophic events that led to the demise of so many species. During the Eocene, approximately 55 to 35 million years ago, the emydid turtles emerged. This is the largest group of extant (contemporary) turtles and the group to which diamondback terrapins belong.

Chelonians, as well as other reptiles, birds, and mammals, are characterized by their ability to produce eggs that have protective coverings supported by extraembryonic membranes. The covering or shell inhibits the egg from drying out and is also supplied with the energy source for the developing embryo. Another important reptilian feature is dry skin or scales. Since these animals can live on land, the scales are an important adaptation to prevent desiccation. The reptile represents the completion of a full transition from water to land-dwelling animal. The ability to produce covered eggs and the ability to prevent desiccation allowed reptiles to break their dependence on the watery habitat utilized by their amphibian ancestors and thus to take advantage of a terrestrial way of life. Although some amphibians can live on land, they must return to the water to lay eggs. Paradoxically, some reptiles, including the aquatic turtles, live in water but must lay eggs on land.

The lack of fossil evidence that would link turtles to an evolutionary precursor leads to much speculation about the evolutionary development of the basic turtle body plan. If we adhere to the classical definition of a reptile, which does not include birds, turtles are the only modern representatives without teeth and they are the only extant vertebrate with a shell. The turtle's shell is a one-of-a-kind evolutionary item. It represents a major anatomical contortion. The shell, which originated from an amalgam of ribs and spine, is fused to the skeleton in such a way that the pelvic and pectoral girdles (hips and shoulders) lie within the rib cage. This unique anatomy makes it difficult to easily propose morphologically based phylogenetic relationships between turtles and other reptiles. The true marvel of this body plan was described by the world-renowned turtle expert Archie Carr (1952):

> The first of the innovations made by the stem reptiles was in a way the most extraordinary and ambitious of all-the most drastic departure from the basic reptile plan ever attempted before or since. By a cryptic series of changes, few of which are illustrated in the fossil record, there evolved a curious and improbable creature which, though it retained the old cotylosaur skull (with no openings in the temporal region), has a horny, toothless beak and a bent and twisted body encased in a bony box the likes of which had never been seen. And more than this, within the box the girdles connecting the legs and the rest of the skeleton had by some legerdemain been uprooted and hauled down to an awkward position underneath the ribs. (p.1)

The presence of a shell undoubtedly had an impact on the status of some turtles in the food web. With a coat of armor, land turtles no longer needed to be swift of foot; they could rely on a unique shielding mechanism for protection from predators. As an added bonus, land turtles no longer needed to be swift to capture their own prey. Although some had dietary preferences, others adopted a vegetarian or generalized omnivorous diet, munching on anything within reasonable reach. Turtle adaptations were apparently successful. With the protection of their shell, turtles could take life at a somewhat more relaxed pace than their evolutionary forebears. And thus they persisted and witnessed the decline of other reptile relatives. The first appearance of the diamondback terrapin from early turtle ancestors remains a mystery. The only known diamondback terrapin fossils, two tiny bones, a nuchal (from the neck region of the shell) and a costal (from the side or lateral portion of the shell) from different individuals, date back to the Pleistocene epoch (approximately

1.65 million to 10,000 years ago). They were found at Edisto Beach, along the coastal plain of South Carolina (Dobie and Jackson, 1979). This region lies within the current range of *Malaclemys terrapin*.

Who would think that turtles are the center of a major controversy among evolutionary biologists? Long considered a living fossil and one of the most primitive reptiles, an egg-laying precursor to snakes, lizards, alligators, and crocodiles, new evidence suggests that turtles may belong to a branch of the evolutionary tree that links them more closely with advanced reptiles and very closely to birds. Biologists have used skull type as one important criterion for determining evolutionary relationships and classification of reptiles. In the anapsid skull, bone completely covers the regions around the eye socket; the skull is solid. This is thought to be the most primitive type of skull, and indeed, turtles are anapsid reptiles. In the diapsid skull, characteristic of reptiles such as crocodiles and also seen in birds and in dinosaur fossils, there are two openings in the skull, posterior to the eye sockets. These openings are believed to be important for attachment of muscles that function in jaw movement. In the traditional evolutionary timeline, anapsid turtles are thought to predate diapsid reptiles.

Other morphological features can be brought into the evolutionary equation. Depending on the methods for comparing morphological characteristics, turtles can be moved up to the top branches or down to the trunk of the evolutionary reptilian tree. Rieppel and Reisz (1999) used powerful computer software to take a fresh look at the fossil evidence, combined with morphological characters seen in reptiles that roam the earth today. Their resulting

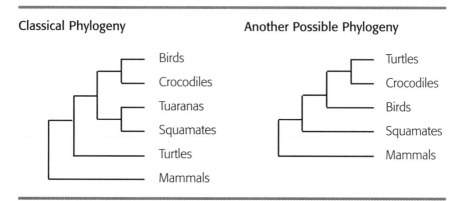

*Fig. 1.1.* *These diagrams depict two theories of turtle evolution.*

analysis led them to hypothesize that turtles are nested within the diapsid clade. Perhaps they were originally diapsid and reverted to the anapsid condition in a later evolutionary transition.

Another method to examine evolutionary relationships employs molecular analysis. DNA sequence information allows scientists to examine evolutionary relationships based on DNA sequence similarities and differences. This technique also relies on powerful computer software to make the necessary analyses and comparisons in DNA sequence between the same genes in different organisms. In many cases, DNA sequence information and morphological comparisons lead to the same conclusions. Hedges and Poling (1999) compared DNA sequences in reptiles using nine genes found within cellular organelles known as mitochondria, and twenty-three genes found within the cell's nucleus, and found strong evidence for a turtle–crocodilian relationship. Therefore, DNA sequence data support the hypothesis that turtles are not as "primitive" or genetically distinct as originally thought. The phylogenetic trees in fig. 1.1 summarize the two theories of turtle evolution.

Terrapins share features of the truly marine turtles as well as their freshwater relatives. But from which ancestor is the terrapin derived: a freshwater turtle or a marine ancestor?

Terrapins belong to the large family of turtles known as the Emydidae. Most members of the family, with the exception of most box turtles (Genus *Terrapene*), are aquatic. Included in this family are some familiar freshwater species such as painted turtles (*Chysemys picta*), spotted turtles (*Clemmys guttata*), wood turtles (*Glyptemys insculpta*), map turtles (*Graptemys*), Blanding's turtles (*Emydoidea blandingii*), cooters (*Pseudemys*), and sliders (*Trachemys*). Among turtles, diamondback terrapins are unique in their habitat and adaptations. The closest turtle relative of the diamondback terrapin is the freshwater map turtle, *Graptemys*. Despite their morphological similarities, there are significant differences in their appearance. But more importantly, they differ physiologically since *Graptemys* is a strictly freshwater turtle while *Malaclemys* is the lone occupant of the brackish water niche. There have been several proposals to explain the evolutionary relationship between the two species, and from a theoretical viewpoint, one could look at current distribution, morphological characters, and physiological adaptations.

Dobie (1981) used detailed analysis of skull and shell features to compare current *Graptemys* with current *Malaclemys*. Osteological (bone structure) comparisons of skull, jaw, and neck and aspects of external morphology led Dobie to propose that *Malaclemys* may have arisen from *Graptemys* or that both arose from a common Eocene precursor approximately 60 million years ago.

Other factors can also provide clues to the origin of *Malaclemys*. In a study of habitat utilization of sympatric reptiles, that is, those that occupy the same territory, in Florida Bay, Dunson and Mazzioti (1989) point to salinity as the limiting factor in the utilization of habitats rich in food and nutrients. Very few reptiles have been able to adapt to a euryhaline environment, i.e., one that has a wide range of salinity, for prolonged periods. They suggest a four-stage evolutionary adaptation to salinity which allows reptiles to regulate the salt content in their body fluids, a process known as osmoregulation, which includes:

1. Behavioral osmoregulation: By keeping the mouth closed except while feeding and drinking rainwater rather than salt water, some freshwater reptiles, such as snapping turtles, can use habitat that is temporarily salty.
2. Physiological specialization: a decrease in net salt uptake, net water loss, and incidental drinking of salt water while feeding.
3. Appearance of salt glands: an extrarenal (non-kidney-mediated) mechanism for elimination of excess salt.
4. Development of larger salt glands and external features suited for pelagic life (the open sea): These traits are seen in marine iguanas and sea turtles. Although the terrapin also has a salt gland, it is different from that of sea turtles and may have evolved independently.

According to this scheme, the diamondback terrapin is an estuarine reptile with an intermediate adaptation to a marine environment. Similar to Dobie's conclusions (1981), this hypothesized evolutionary pathway also suggests that *Malaclemys* may have originated from *Graptemys* or a common freshwater ancestor.

In agreement with this interpretation, molecular evidence points to a fresh to brackish water evolutionary pathway for *Malaclemys*. Lamb and Osentoski (1997) used molecular data to propose an evolutionary relationship between *Malaclemys* and *Graptemys*. They focused on certain mitochondrial genes that are often used in evolutionary and phylogenetic analysis because of their tendency to be quite variable. The general assumption behind such an approach is that the more similar the DNA sequences of specific genes, the more similar the species that are being compared. When the variable mitochondrial control region and the more conserved cytochrome *b* gene were sequenced and compared, the genetic data point to a scenario in which both genera evolved from a common ancestor some 7 to 11 million years ago during the late Miocene.

It is certain that diamondback terrapins have been around for a long time. As Wood (1977) stated:

> In view of the fact that diamondbacks have no apparent competitors in the salt marshes to which they are uniquely adapted, that this habitat may be of considerable antiquity, that they are quite different from all the emydines except *Graptemys*, and that emydines are a fairly ancient group (being known from late Paleocene and early Eocene deposits of Western North America), *Malaclemys* may be a taxon that has persisted over a fairly great time span while undergoing little change. (p. 420)

## TAXONOMY

Even without the evolutionary pieces falling into place, it is still possible to classify the diamondback terrapin based on morphological traits and current geographic distribution (Ernst and Bury, 1982). With a general turtle phylogeny in mind, the diamondback terrapin has been placed as follows:

**Kingdom:** Animalia (i.e., animals).

**Subkingdom:** Eumetazoa (animals having definite symmetry and tissues).

**Phylum:** Chordata (chordates have the following four characteristics: a hollow dorsal nerve cord, a notocord, pharyngeal slits, and a postanal tail, at some point in their development).

**Subphylum:** Vertebrata (vertebrates are basically chordates with a spine).

**Class:** Reptilia (from the Latin, creepy, crawly).

**Order:** Chelonia (from the Greek word suggesting interlocking shields or armor).

**Family:** Emydiiae (a freshwater turtle, in Aristotle's "History of Animals").

**Genus:** *Malaclemys* (from the Greek, mollusk-eating turtle).

**Species:** *terrapin* (from Algonquian, edible turtle).

**Subspecies:** *terrapin; centrata* (from the Greek, kentron, center, refers to formation of growth rings on scutes); *tequesta* (after Tequesta, Native American tribe in eastern Florida); *rhizophorarum* (Greek for genus of mangrove in the habitat of this subspecies); *macrospilota* (from the Greek macron, large and spilados, spot; refers to the yellow spot at the center of each scute); *pileata* (from the Latin, capped; refers to black marking at top of the head); and *littoralis* (from the Latin, littoralis, seashore).

The morphological characteristics of the diamondback terrapin sub-species are outlined in table 1.1.

**Table 1.1** Morphological Comparison of Diamondback Terrapin Subspecies

| Subspecies | Common name | Distinguishing morphological features | Range |
|---|---|---|---|
| *terrapin* (Schoepff, 1793) | Northern diamondback | No knobs on median keel; carapace is black to light brown/olive with distinct concentric rings on scutes; plastron is light-colored, yellow, orange, or greenish gray; carapace is wider behind bridge | Cape Cod, Mass., to Cape Hatteras, N.C. |
| *centrata* (Latreille, 1802) | Carolina terrapin or Southern diamondback | No knobs on median keel; posterior margins curled upward | Cape Hatteras, N.C., to Northern Florida |
| *tequesta* (Schwartz, 1955) | Florida East Coast terrapin | Median keel has posterior-facing knobs; carapace dark or sometimes tan with light centers on scutes; no pattern of light concentric circles | Florida's east coast |
| *rhizophorarum* (Fowler, 1906) | Mangrove terrapin | Median keel has bulbous knobs; oblong shell; carapace is brown or black; plastral scutes are outlined in black; neck and forelimbs are uniform gray with no markings; black striations may be found on hindlimbs | Florida Keys |
| *macrospilota* (Hay, 1904) | Ornate diamondback | Median keel has terminal knobs; carapace scutes have orange or yellow centers | Florida Bay to Florida Panhandle |
| pileata (Wied-Neuwied, 1865) | Mississippi diamondback, Biloxi terrapin, Gulf terrapin | Median keel has terminal tuberculate knobs; plastron is yellow; upturned edges of marginals are yellow; dorsal surfaces of head, neck, and limbs are dark brown or black | Florida Panhandle to western Louisiana |
| littoralis (Hay, 1904) | Texas diamondback | Deep carapace with terminal knobs on median keel; plastron is very pale; dorsal surface of head is white or light color | Western Louisiana to Western Texas |

*Source:* Adapted from Carr (1952) and Ernst, Lovich and Barbour (1994).

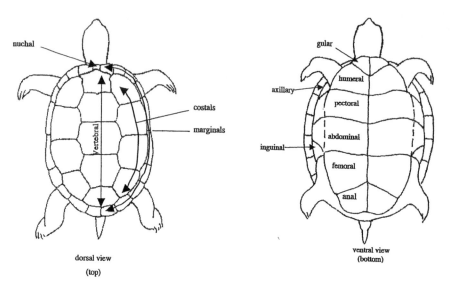

Fig. 1.2. Diagram of the scute pattern on carapace (dorsal view) and plaston (ventral view).

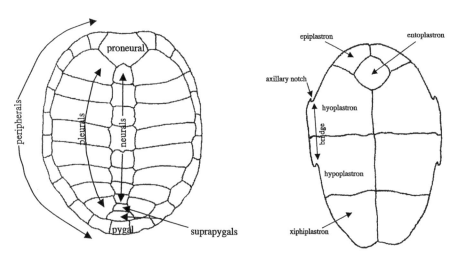

Fig. 1.3. Arrangement of turtle internal bony structure; bones that would be visible if the scutes of the carapace and plastron were removed.

## Anatomy and Morphology

### The Turtle

*The turtle lives 'twixt plated decks*
*Which practically conceal its sex*
*I think it clever of the turtle*
*In such a fix to be so fertile*
—*Ogden Nash*

### THE SHELL

When we think about turtle anatomy, the first thing that comes to mind is the shell. The shell is made of bone. The upper shell is called the carapace, and during embryonic development it is formed by the fusion of the spine with bones that would normally form the rib cage in other animals. The ribs of other animals are present to protect vital organs, but in turtles the ribs serve as buttresses to support the carapace. Turtles do not have a breastbone or sternum. The plastron or bottom shell is composed of bone and has no equivalent structure in other reptiles: its embryological origin remains a mystery. An unconfirmed study suggests that the plastron may originate from an embryonic area known as the neural crest, a region that gives rise to muscles, blood vessels, and facial bones (Pennisi, 2004). The bridge is the bony segment that connects the two halves of the shell and serves as a brace or structural support that prevents the upper and lower shells from collapsing upon one another after a heavy impact. The plastron of the diamondback terrapin does not have a hinge and is thus immobile; the terrapin cannot completely hide within, as can some turtles. The shell is more than a protective structure that simply covers a reptilian body. It is an integrated, much modified part of the body. Fat, stored under the shell, imparts the characteristic flavor to turtle soup. Due to the development and placement of the shell, the turtle has a body plan that is inside-out. As mentioned earlier, its pectoral and pelvic girdles (shoulders and hips) are inside its ribs.

The shell of most turtles has a similar structure (see Carr, 1952; Alderton, 1988). A cross section of the turtle shell would reveal a two-layered structure. The inner, dermal layer is composed of fused bony plates, while the outer, epidermal layer is scaly tissue, filled with keratin, a connective-tissue protein found in the hair, nails, hooves, and horns of other animals. The epidermal layer of the carapace and plastron is divided into segments called "scutes" that form a mosaic over the bones. In diamondback terrapins the scutes do not overlap (fig. 1.2). Instead, they abut one another like closely spaced tiles. The pattern of scutes does not correspond to the distribution and pattern of the underlying dermal bones (fig. 1.3). There are many more dermal bones than the scutes that cover them. In the diamondback terrapin carapace, we can observe vertebral scutes along the midline or center of the back, pleural or costal scutes along each side, one cervical or nuchal scute near the neck region, and an apron of marginal scutes. The plastron has scutes that are divided into gular, humeral, pectoral, abdominal, femoral, and anal sections (fig. 1.2). It is not unusual to find diamondback terrapins with minor scute variations. Some have extra, missing or sectored scutes (see fig. 1.4). These minor anomalies arise during shell development and do not compromise the viability of the terrapin.

The main ridge along the midline of the carapace is the keel. In the diamondback terrapin, it is sometimes smooth and barely noticeable, but it may also be prominent with protruding knobs, especially in younger animals and also among some of the subspecies (fig. 1.5). Anomalies are sometimes seen in the number and formation of ridges that make up the keel. Some ridges may be subdivided or slightly deformed. The major vertebrae that make up the turtle spine are fused to the carapace and are therefore very rigid. The smaller tail and neck vertebrae are covered by muscle and have great flexibility.

As turtles grow, so do their shells. Under each scute, within the epidermal layer, Malphigian cells produce keratin. Scutes grow outward in all directions from a central section known as the areolus. As the scute expands, keratin is added. Pigment is also deposited to produce spotting patterns or rays, unique to each scute. Diamondback terrapins are only active for part of the year, and growth is restricted to these periods. As a result, there will be a new margin of growth that will represent a season of eating and activity (fig. 1.6). Sometimes it is possible to estimate the age of diamondback terrapins by counting the growth rings, or annuli, on carapace or plastron, similar to counting the growth rings of tree cross-sections to estimate the age of a tree. As terrapins age, the growth rings may become worn, smooth and difficult to discern. After six to eight years, when growth occurs in smaller increments, it is very difficult to distinguish annuli at the margins of the scutes. Terrapins raised indoors, in captivity, do not have a hibernation period and often experience continuous growth, independent of the seasons. There may be more gradual, rather than distinct, deposition of keratin during enlargement of scutes.

Terrapins have been found with shell injuries resulting from boating accidents, dredging operations, and close encounters between nesting females and automobiles. If the injuries are not life-threatening, healing will occur and the

*Fig. 1.4. Four anomalous scute patterns.*

*Fig. 1.5. The keel is particularly pronounced on hatchlings.*

*Fig. 1.6. Growth rings or annuli are easy to discern in young terrapins.*

bone will gradually mend. Wildlife veterinarians have developed treatments for terrapins with more seriously injured shells. Temporary patching of the shell with fiberglasslike materials or special taping compounds makes it possible to preserve proper dermal bone alignment so that natural healing can take place. Healing is a slow process, and it may take over a year for the bones of the shell to mend.

Terrapin number 1195 presents a case study in shell healing. This 13-year-old female was found on the beach near Wellfleet Harbor on Cape Cod in mid October, 2001. Sediments were being removed from the shallow harbor, and it seems likely that number 1195 was a victim of heavy dredging equipment. Her carapace was cracked, but the underlying vital organs were not penetrated. In addition, her left bridge was broken. After she was brought to the attention of the Wellfleet Bay Wildlife Sanctuary, Don Lewis, terrapin researcher and Massachusetts Audubon Society volunteer, transported the bleeding terrapin to the Cape Wildlife Center in Barnstable, Massachusetts, where veterinarians Rachel Blackmer and Catherine Brown began treatment. The broken bridge was diagnosed as the more serious injury. The bridge bones were taped into place and the terrapin was kept in a heated tank over the winter so that her progress could be monitored. Lewis provided a daily clam and oyster feast for number 1195. By springtime, her bridge was completely healed. Her return to Wellfleet Harbor was a celebrated event that was covered by the local newspaper, *The Cape Codder,* on April 19, 2002.

At Cape May, along the southern New Jersey shore, approximately 500 female terrapins each year are struck by automobiles. Most of these are killed, but occasionally, an auto victim will survive. Such a survivor will have her shell fiberglassed and will spend the fall, winter, and spring at Richard Stockton College of New Jersey under the watchful eyes of Drs. Roger Wood and Roz Herlands. These females are returned to the Cape May salt marshes after sufficient healing has occurred (fig. 1.7).

Although life in a shell is a successful strategy for turtles, it is not without its disadvantages. Wherever a turtle goes, the heavy shell must go with it. This may be less of a problem for aquatic turtles than terrestrial species, but the female diamondback terrapin, as well as her hatchlings, must sometimes travel about on land. Their limb muscles operate very differently from those of nonshelled vertebrates. Terrapins can move relatively rapidly on land but are not often quick enough to elude a terrestrial predator or a determined researcher.

*Fig. 1.7.* Roger Wood holds a female whose carapace was broken in an auto accident. The shell has been treated by applying a fiberglass-type patch to allow healing.

## INTERNAL ANATOMY

Within the terrapin shell lie structures that are very similar to those of other turtles. Details of turtle internal anatomy have been reviewed by others (Pope, 1946; Carr, 1952; Alderton, 1988; Ferri, 2002). I describe the most significant aspects.

The cardiopulmonary system (heart, blood vessels, and lungs) of turtles is very specialized. Aquatic turtles are air breathers but can spend months submerged during hibernation. They have great tolerance to anoxia (lack of oxygen) and can undergo long periods without breathing. They also differ markedly from other vertebrates that are able to use chest movement via muscles of the thorax and a diaphragm to fill and empty their lungs. Aquatic turtles are thought to use muscles at the base of their limbs and beneath their vital internal organs to assist them in breathing. To inhale, these muscles contract, the body cavity or coelom enlarges, and pressure is reduced, thus allowing the lungs to fill. To exhale, the muscles relax and water pressure will do the rest in order to push the organs against the lungs and allow air to be forced out (Carr, 1952). The volume of air in an aquatic turtle's lungs will affect its buoyancy. Similar to other turtles, the terrapin has a three-chambered heart constructed of two atria and one ventricle. The hint of structural division in the ventricle provides a preview of the development of the four-chambered heart found in crocodilians, birds, and mammals (Carr, 1952; Ferri, 2002). The terrapin digestive and excretory systems resemble those of other freshwater emydid turtles (Carr, 1952; Ferri, 2002). Terrapins do not have a rectum; the cloaca (from the Greek word for "sewer") serves as an all-inclusive excretory and genital area, collecting fecal matter from the gut and urine from the bladder and serving as the location for the genital organs. The bladder is often emptied when turtles are handled—a response that most researchers become wary of after their initial "christening."

The turtle brain is well developed in those regions that process visual and olfactory signals. Turtles have well-developed eyes, protected by heavy lids, and they have a good sense of smell, but they may not be well equipped for hearing. Their tympanic membrane, behind their eyes, is covered with skin, lacks an external opening, and may only respond to low notes (Alderton, 1988). It is clear that terrapins can respond to some sounds. In the absence of any visual stimulus, farm-raised terrapins, kept in pens in the early 1900s, became extremely responsive in anticipation of feeding when they heard their food being chopped up (Coker, 1906).

Within the brain, turtles have a relatively large pineal body, an area that produces the hormone melatonin. In many animals, this section of the brain responds to an internal biological clock that governs daily (circadian) rhythms. The clock ticks in approximately twenty-four-hour cycles to respond to external cues such as daylight. Turtles also have other internal timepieces. A field study of a Long Island, New York, terrapin cluster suggests that *M. t. terrapin* has activity patterns that are regulated by tides. Swimming was correlated with high tides, while basking was correlated with high and low tides, depending on the brightest hours of the day. The idea that this was internally regulated and represented an innate tidal activity cycle stemmed from observations of laboratory-raised hatchlings that displayed the same types of approximately six-hour daily fluctuations in locomotor activities (Muehlbauer, 1987). It is highly probable that terrapins also have another clock that governs yearly (circannual) behaviors such as mating, winter dormancy (hibernation or brumation), and local movement. But very little is known about the internal mechanisms that drive these longer cycle periods.

**Table 1.2**   Mean Adult Size: Size and Age at Maturity
of Diamondback Terrapins

| Geographic Location | Linear Measurements in cm (in.) Correspond to Straight-Line Plastron Length | |
| | Males | Females |
| --- | --- | --- |
| Wellfleet, Cape Cod, Massachusetts *M. t. terrapin* (Wellfleet Bay Wildlife Sanctuary) | 9.7 cm (3.8 in.); 263 g (0.58 lb.) average (largest = 400g [0.88 lb.]); Maturation: reach 8 cm (3.15 in.) by 5th year. | 16.4 cm (6.46 in.); 1063 g (2.34 lbs.) average Maturation: 8–10 years; 14 cm (5.51 in.) |
| Rhode Island *M. t. terrapin* (Goodwin, 1994) | – | 20.0 cm (7.87 in.); range = 17.5–22.5 cm (6.9–8.9 in.) |
| Little Beach Island, New Jersey *M. t. terrapin* (Montevecchi and Burger, 1975) | – | 15.4 cm (6 in.) average; range = 13.2–18.4 cm (5.2–7.24 in.) |
| Patuxent River, Maryland: *M. t. terrapin* (Roosenburg, 1994, 1996) | Maturation: 300 g (0.66 lbs.); 4–7 years. | Maturation: 1100 g (2.43 lbs.); 17.5 cm (6.9 in.); 8–13 years. |
| North Carolina: *M.t. terrapin x centrata* (Hildebrand, 1932) | Largest = 12 cm (4.72 in.) Maturation: 8–9 cm (3.2–3.5 in.); 5 years. | Largest =18.5 cm (7.3 in.). Maturation: over 13.7 cm. (5.4 in.) Smallest = 11.97cm (4.7 in.), 7 years; (range = 4–8 years). |
| South Carolina *M.t. centrata* (Gibbons et al, 2001) | 10.26 cm (4 in.) average; 242 g (0.53 lb.) Maturation: 9 cm (3.5 in.); 3–4 years. | 14.42 cm (5.7 in.) average; 667 g (1.47 lbs.). Maturation: 13.8 cm (5.4 in.) 6–7 years. |
| Merritt Island, Florida *M. t. tequesta* (Seigel, 1984) | 10.4 cm (4 in.) average; 283 g (0.62 lb.) Maturation: over 9.5 cm (3.7 in.); 2–3 years. | 15.4 cm (6 in.) average; 886 g (2 lbs.) Maturation: over 14 cm (5.5 in.) 4–5 years. |
| Louisiana *M.t. pileata x littoralis* (Cagle, 1952) | 3 years | 6 years |
| Texas *M.t. littoralis* (Unpublished report by K. A. Holbrook and L. F. Elliot, 1997) | 12.6 cm (5 in.); range = 11.1–15.3 cm (4.4–6.0 in.) | 16.1 cm (6.3 in.); range = 10.1–22 cm (4.0–8.7 in.) |

### EXTERNAL MORPHOLOGY: SEXUAL DIMORPHISM

Males and females of many animal species can be readily distinguished from one another by observable differences in external morphology that are called secondary sexual characteristics. Sexual dimorphism in the diamondback terrapin can be seen in adults at the level of total body size. Carr (1952) observed that adult diamondback terrapins have greater size disparity between the males and females of the species than any other North American turtle. Although the size of terrapins generally varies and is normally distributed, similar to the different sizes and shapes of humans and other animals, adult females are always much larger than males. For example, on Cape Cod, females often attain plastron lengths (PL) (straight-line measurements with a caliper) of 16 centimeters (about 6 inches), while adult males average around 10 centimeters (about 4 inches) PL. Adult females may weigh 1000 grams (2.2 pounds) or more, while adult males top out at an average of approximately 275 grams (0.6 pounds). This size disparity exists among terrapins throughout their range (table 1.2).

Adult males and females are also distinguishable by head size and the size and shape of the tail. Mature females have larger heads than mature males; the neck muscles in both sexes are well developed to allow rapid retraction of the head into the shell when the terrapin is threatened.

The tail of the female is shorter and narrower than that of the male (plate 1). The tail of the diamondback terrapin, similar to that of other aquatic turtles, is muscular and flexible and can serve as a rudder for steering. For males, it is important in grasping and aligning with females during mating.

The cloacal opening of the male is posterior to the shell margin, while the opening of the female does not extend far beyond the apron of the shell. The carapace of the adult male is flatter than that of the female. Although sexual dimorphism is seen in adult terrapins, it is more difficult, if not impossible, to distinguish the gender of hatchlings and juveniles by external observation.

### JAW AND APPENDAGES

Similar to other emydid turtles, the limbs of the diamondback terrapin have a structure that is intermediate between a true sea turtle, whose limbs are basically flattened and shaped like flippers, and the thick and columnar limbs of land-dwelling turtles. Terrapin limbs terminate with webbed toes and sharp nails; rear limbs are larger and more powerful than forelimbs. The webbing is a feature of most aquatic turtles and serves the terrapin well for life in the water. Unlike sea turtles, which can glide very swiftly through the water using up-and-down, winglike movements of their front flippers, diamondback terrapins rely on swimming strokes similar to those of their freshwater emydid cousins, employing alternate diagonal movement of front and rear limbs. Nails on the webbed toes are employed for digging into sand, soil, or mud, as when females dig nests or when terrapins burrow into the muddy bottoms of creeks and marshes during the cold weather. Nails also assist terrapins in grasping food while they use their powerful jaws to tear food into bite-sized pieces. Diamondback terrapins are surprisingly good climbers. With the help of their nails, they have been known to climb vertical surfaces. During terrapin farming attempts in Beaufort, North Carolina, they were observed to scale several feet up the walls of their pens in attempts to escape (Coker, 1906).

The terrapin upper jaw is usually thick and white, giving the appearance of lips shaped into a comical smile (fig. 1.8). Males as well as females occasionally have a dark coloration to the upper jaw, resembling a moustache. The terrapin jaw is a hard bony hinge without teeth. It is capable of crushing shells of mollusks such as snails and crustaceans such as crabs. These jaws can draw blood from a researcher who is not vigilant. One unfortunate encounter with a terrapin demonstrates both their climbing ability and their strong jaws: Jack Rudloe was working as a specimen collector in Florida's Panacea Channel. On one of his trips, many of the terrapin specimens climbed out of the buckets in which they were contained and were observed crawling over the boat's deck. It wasn't long before a large female clamped her jaws around Rudloe's big toe and wouldn't let go until he submerged his foot in a bucket of water (the same method used to induce blue crabs to release their grip). Rudloe's toe remained bruised and swollen for days afterwards (Rudloe, 1979).

## Geographic Variation

Even amateur herpetologists can identify a diamondback terrapin and pick it out from a lineup of similar turtles. Aside from habitat preferences, there are morphological characteristics that are unique to diamondback terrapins such as their distinctive shell patterns and the various designs of spots and stripes on their exposed integument. They also differ quite a bit from each other: no two are exactly alike. The grouping of diamondback terrapins into subspecies is based primarily on overall visible characteristics and geographic location.

Although all members of *Malaclemys terrapin* have similar morphology, experienced researchers have characterized seven subspecies. The shape of the

*Fig. 1.8. The upper jaw of the diamondback terrapin is thick and strong. When it is lightly colored it resembles clown lips. When it is darkly colored it resembles a moustache.*

shell varies slightly, and there are also subtle variations in coloring and patterning (plate 2). These morphological differences among the subspecies were summarized in table 1.1. I have observed much overlap in these categories. For example, black striations on hind limbs, sometimes referred to as "striped pants," are descriptive of the mangrove terrapin, but not all mangrove terrapins have black stripes on their limbs. Furthermore, I have seen these distinctive markings on New York terrapins.

There is also some size variation among the subspecies (table 1.2). The Chesapeake population has larger adult specimens than the more northern and southern populations. Rhode Island females are larger than Massachusetts and New Jersey counterparts. The *M. t. centrata* females from South Carolina are the smallest.

Coloration is mediated by the pigment melanin, which is deposited in various amounts and in a variety of patterns, on both shell and skin. Carapace color varies from pale olive green to almost black, with many shades of greenish-gray and brownish-green in between. The plastron is lighter in color than the carapace and can be pale yellow to bright orange. Plastron coloration tends to become duller as terrapins age. Each hatchling possesses distinctive plastron spotting patterns or markings that persist for several years but eventually become blurred as the terrapin grows. We have found these markings to be unique to each hatchling and have used the plastron patterns as a convenient way to identify young terrapins (plate 3). In addition to shell color variation, skin color and skin markings differ among subspecies (table 1.1).

Some of the subspecies are found in wide geographic tracts that span two biogeographic regions (Fig. 1.9). *Malaclemys t. terrapin* ranges from Cape Cod to North Carolina; *M. t. centrata* is found from South Carolina to northern Florida. Florida has the highest number of subspecies; five of the seven subspecies can be found in the Sunshine State. In 1940, when Archie Carr summarized the fauna of Florida, only three subspecies were listed and described (Carr, 1940). Although a specimen of an apparently different subspecies, *M. t. rhizophorarum*, was discovered in 1904, the subspecies was not officially recognized until several years later (Carr, 1946, 1952). Roger Wood provided evidence for the spatial distinction of the two most southern Florida populations. *Malaclemys t. rhizophorarum* is only found on small islands south of Marathon Key, most of which are located in the Key West National Wildlife Refuge. This population may be morphologically and geographically distinct from the group that occupies Florida Bay (Wood, 1992).

Recent genetic evidence questions whether the current subspecies classification is valid and accurately reflects the population distribution of terrapins. For example, molecular studies point to a very close genetic relationship between all subspecies of Florida terrapins. Furthermore, genetic analysis, based on mitochondrial genes, shows divergence between the mid-Atlantic (*centrata, terrapin*) and Gulf Coast (*rhizophorarum, macrospilota,* and *pileata*) subspecies (Lamb and Osentoski, 1997). This divergence was originally noted by Carr on the basis of morphology (Carr, 1946). This type of phylogenetic

*Fig. 1.9.* Geographic range and subspecies distribution for Malaclemys terrapin.

split between Atlantic and Gulf terrapins may have been caused by landmass expansion believed to have occurred approximately 15,000 years ago as a result of a massive sea-level drop. Along with other factors, changing sea level created an ecological and spatial barrier between the two populations. Finer scale genetic studies may lead to reevaluation of the subspecies designations. Population genetic studies and their implications are further described in chapter 6.

Occasionally, at the intersection of the range of two subspecies, terrapins are found crossing over to the range of another subspecies. If mating occurs, intergrades or hybrids will be produced. These intergrades have been observed on the west coast of Florida, where *M. t. rhizophorarum* has been found trespassing in the range of *M. t. macrospilota*, near Naples (Johnson, 1952). The subspecies can interbreed. Some attempts at increasing market value for terrapins at the beginning of the twentieth century included the production of hybrids: *terrapin* x *centrata*. Ecological differences among the subspecies are highlighted in chapter 2.

## Life in a Salty Environment

Salt, in the form of sodium chloride (NaCl), is a compound that is necessary for life, but there are instances when there may be too little or too much of a good thing. Many small creatures that live in salt water, such as invertebrates, exist in a situation in which the salt concentration within their body fluids is exactly the same as the salt concentration in their environment. Turtles, on the other hand, belong to the group of animals that are able to regulate the salt concentration in their blood and body fluids. The salt concentration will be relatively constant despite the salt or lack of salt in their environment and will generally be about one-third that of seawater. This type of regulation is known as osmoregulation and is not always precise. Under certain circumstances, too much water may be lost and the internal salt concentration may rise to a higher than ideal level. This causes dehydration that may be serious if the condition persists.

Osmoregulation is a particular challenge for diamondback terrapins because the salinity of the diamondback terrapin habitat is quite variable. In some river estuaries where considerable mixing of fresh and salt water occurs, the salt concentration may be relatively low, less than 10 parts per thousand (ppt). Waters that flow across marshes and in tidal creeks may have an intermediate salinity that changes after rainfall and with tides (10 to 20 ppt). Further out into the larger embayments, salinity can increase to 20 to 30 ppt and can even approximate that found in marine environments of the open seas (30 to 35 ppt). Captive terrapins can survive and grow under a variety of salinity conditions and even in fresh water. The diamondback terrapin adjusts to these varied salinities and can spend considerable time in fresh water, brackish water, or water with marine salinity levels.

Most turtles cannot survive very long in 100 percent seawater. The diamondback terrapin is the only emydid turtle that can spend weeks at a time under such conditions. Freshwater emydid turtles that are kept in seawater become salt loaded and osmotically dehydrated. Some, like the snapping turtle, can tolerate brief excursions into brackish water but are not truly adapted to life in an estuary or tidal creek. Truly marine turtles have a special adaptation: a type of orbital or eye gland, called the lacrymal gland, which works with the kidneys to prevent sea turtles from dehydration. Sea turtles effectively desalinate the water in their environment. Salt is excreted and seawater is converted to fresh water. These marine turtles never drink fresh water. Unlike sea turtles, the diamondback terrapin is dependent on an external source of fresh water. Clues about the importance of fresh water to terrapin

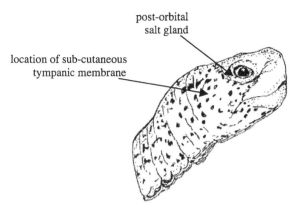

post-orbital
salt gland

location of sub-cutaneous
tympanic membrane

*Fig. 1.10. Terrapin salt gland is located behind the eye. The position of the tympanic membrane is indicated. Neither structure is visible in live specimens.*

health were gleaned in the early 1900s during attempts to farm raise terrapins for the food trade (Coker, 1906). Trappers who impounded them for long periods of time while waiting for better market prices recognized the value of periodic hosing with fresh water to maintain the weight and viability of their captives. In its natural environment, the diamondback terrapin copes with salt in its surroundings in a number of physiological and behavioral ways.

Physiological adaptations to salinity include a pair of orbital glands, similar to those of sea turtles. While terrapin and sea turtle salt glands are similar, they may have evolved independently. Initial studies of the location and function of the terrapin salt gland produced some confusing results. Salty tears, collected from the orbital region of terrapins, were initially attributed to the secretions of the Hardarian gland. It was latter shown that the Hardarian gland was similar to that of freshwater turtles and had a function in the secretion of organic compounds (Cowan, 1971). Ultrastructural studies utilizing staining and electron microscopy allowed visualization of a postorbital (behind the eye) gland in terrapins (lacrymal gland) that is considerably different from that of freshwater turtles and has a structure that implicates its role in salt secretion (fig. 1.10) (Cowan, 1971, 1973; Dunson, 1976).

The lacrymal glands can work as an accessory kidney, enabling terrapins to produce tearlike secretions with high salt content. However, these glands may not be as efficient at eliminating excess salt as their counterparts in sea turtles. Terrapins kept in seawater always have measurably higher sodium levels and higher plasma osmotic pressure than terrapins kept in fresh water.

Dunson and colleagues performed a series of osmoregulation studies on diamondback terrapins in the 1970s (Dunson, 1970; Dunson and Dunson, 1975; Robinson and Dunson, 1976). Terrapins were maintained under various conditions: fresh water, salt water, and brackish water of varying salt concentrations. Immersion of terrapins in salt water was shown to activate the lacrymal gland, but the total secretory capacity is low compared to the salt glands of marine reptiles and the lacrymal glands of sea turtles (Dunson, 1970). Furthermore, the terrapin lacrymal gland may only be activated during prolonged dehydration. During laboratory tests, it was necessary to artificially salt load terrapins to observe a high level of activation of the enzymatic activity responsible for sodium efflux (Dunson and Dunson, 1975). Thus, it appears that the lacrymal gland activity of terrapins is modest and cannot operate alone to routinely prevent salt concentrations in blood and body fluids from increasing. The gland is not active enough to allow for complete osmoregulation in 100 percent seawater. To further complicate our understanding of osmoregulation, it appears that terrapins can inhabit seawater for many months without even utilizing the function of the lacrymal gland (Dunson and Mazzoti, 1989). Some studies suggest that severe dehydration must set in before the lacrymal gland is maximally active (Dunson and Dunson, 1975).

But there are additional mechanisms at work. The diamondback terrapin also has a relatively impermeable integument. This means that the turtle's skin and tissues have a low permeability to salts. In addition, there is a low permeability to water, which helps to prevent diuresis (loss of water) and thus keeps the volume of water in terrapin tissues at a high level (Dunson, 1970; Robinson and Dunson, 1976). Terrapins kept in fresh water weigh up to twice as much as terrapins of the same plastron length that were kept in salt water (Robinson and Dunson, 1976). The weight difference can be attributed to water uptake. Diamondback terrapins that have been kept in seawater hydrate rapidly and dramatically when exposed to fresh or low saline water. A reduced urinary output via the cloaca may help to prevent dehydration when diamondback terrapins are in seawater. Another important physiological adjustment to a salty environment is the tolerance of the diamondback terrapin to functioning with concentrated body fluids. As mentioned earlier, there can be a considerably higher sodium concentration in blood and orbital fluid, depending on whether the terrapin is in fresh or salt water.

Gilles-Baillen (1970, 1973b, 1973c) proposed a mechanism for osmoregulation in which urea, retained in the bladder as a result of decreased urinary output, spills into the blood and contributes significantly to the increased osmotic blood pressure of terrapins that remain in 100 percent seawater for long periods of time. Thus, one of the waste products of nitrogen metabolism can be harnessed to assist in osmoregulation.

When Robinson and Dunson (1976) studied the overall rate of sodium and water exchange in terrapins they determined that terrapin skin is impermeable to sodium and that almost 100 percent of sodium uptake was via an oral route. Terrapins are not able to hydrate when salinity is greater than 20.4 ppt. When fresh water is available, it is likely that terrapins have a system for rapid intake of fresh water and its expedited storage in subcutaneous tissues. The amount of water taken in by terrapins was shown to be dependent on the salinity of the water. Salt-loaded animals did not drink when salinity was 27.2 to 34 ppt; they drank slightly when salinity was in the range of 13.6 to 20 ppt, and they drank copiously at salinities between 0 and 10.2 ppt. These salt-loaded terrapins were able to completely hydrate within fifteen minutes of drinking fresh water (Davenport and Macedo, 1990).

When terrapins are reared in fresh water, some develop swellings at the base of their appendages, which disappear when they are exposed to salt water for extended periods. These "water bags" are storage compartments and reflect the ability of the terrapin to store interstitial water (water in a type of extracellular fluid compartment) (Robinson and Dunson, 1976). These swellings are also seen in terrapins in the wild, where they presumably have a similar water storage function. They allow terrapins to fill up with fresh water so that these stores can be tapped when conditions become more saline.

In addition to physiological adjustments to a saline environment, researchers have documented several behavioral responses that contribute to the overall salt tolerance of the terrapin. Diamondback terrapins take advantage of any occurrence of fresh water in their surroundings. The infusion of fresh water provided by rainfall causes them to drink copiously and hydrate rapidly. In a laboratory setting, simulated rainfall, produced by a watering can, caused diamondback terrapins to come to the surface to drink. The rapid response to the simulated rain appeared to be triggered by vibrational or visual cues or a combination of both types of stimuli (Davenport and Macedo, 1990).

Terrapins kept in 100 percent seawater were shown to drink fresh water from the transient films that form on the water surface before mixing occurs. These films can be as thin as 1.6 millimeters (0.06 inches). When the films are thin, terrapins assume a characteristic drinking posture in which they arch their necks in order to bring their mouths to the same level as the films. Some terrapins drank from water pockets found in the upturned marginal scutes or

*Fig. 1.11. Basking promotes the drying out of the shell and the shedding of old keratin.*

limb pockets of other terrapins or even their own front limb pockets (Bels, Davenport and Renous, 1995). Some enterprising terrapins simply opened their mouths at the water's surface to catch the simulated rain. This is an important survival technique that allows terrapins to rapidly hydrate in habitats in which fresh water may be a scarce resource.

The osmoregulatory studies just described suggest that in their natural habitats terrapins will drink significant amounts of fresh water when it is available and that they are able to take advantage of rainfall at the surface of water and on mudflats. Diamondback terrapins may also utilize additional behavioral approaches to osmoregulation. Terrapins have been shown to alter their food intake and basking behavior if fresh water is in short supply. The diet of adult terrapins is expected to have the same salt content, that is, be iso-osmotic, with the terrapin's environment. Since the terrapin can only eat underwater, some seawater will also be taken in during ingestion of food. Davenport and Ward (1993) showed that food intake amounted to an average of 7.2 percent of body weight when terrapins were fed in fresh water. In contrast, food intake decreased significantly in salt water. Thus, terrapins appear to exhibit hyperphagia (eating of large quantities of food) in fresh water to maximize energy intake, while at the same time minimizing incidental salt intake during ingestion of food.

Since reptiles are ectothermic (commonly referred to as cold-blooded), they often resort to regulating their body temperature by basking. In sunlight, basking, technically known as emersion, functions in thermoregulation. Basking can increase body temperature to maximize physiological processes such as digestion, but it is not without a down side. Basking exposes turtles and increases the likelihood of their detection by a predator. However, Davenport and Magill (1996) presented evidence suggesting that basking may be another behavioral aspect of the osmoregulatory mechanism in terrapins. They observed a progressive increase in basking frequency as a function of the length of time of deprivation of fresh water. When fresh water became available, there was a rapid decline in basking frequency. The first impression one would have about the observed increase in basking behavior with freshwater deprivation is that terrapins will lose even more water and undergo dehydration. Indeed, a significant water loss occurs. But concomitant with water loss, basking results in a decreased salt influx while salt efflux continues. This prevents further concentration of body fluids. Consequently, for terrapins, basking may function not only in thermoregulation; it may also have a role in osmoregulation. Thus, for diamondback terrapins, adjustment to life in a salty environment is a complex issue of osmoregulation involving a suite of physiological and behavioral mechanisms.

Although juvenile and adult terrapins have apparently mastered osmoregulation in an environment that experiences fluctuating levels of salinity, younger terrapins, including hatchlings and yearlings, may have a more difficult time coping with higher salinities. A number of tests performed under laboratory conditions indicate that hatchlings have a lower salt tolerance than adults. This may seem surprising when we consider that many terrapin nesting sites are adjacent to areas in which the nearest water body has a salinity approaching or equaling that of seawater. How do the hatchlings survive in this seemingly hostile osmotic environment?

Dunson (1985) completed a physiological study of sodium and water influx and efflux in adults and hatchlings. He found that hatchlings do not grow well in 100 percent seawater until they have achieved a weight of approximately 50 grams (0.11 pounds). Optimal growth was observed in 25 percent seawater (approximately 8 ppt), with slower growth at higher and lower salinities. In the laboratory, headstarted hatchlings can be pumped up to 50 grams (0.11 pounds) in a matter of months. But for most hatchlings in their natural habitat, this weight will not be achieved until the terrapin is from one to three years old. Dunson also found that hatchlings have a fully functional salt gland but they still have difficulty adjusting to a salty environment. For all terrapins, the rate of sodium influx was related to the salt content of the feeding water, not to the amount of food eaten. This suggests that one of the major sources of sodium influx in hatchlings, as for adults, is incidental water intake during feeding, rather than the food itself. Dunson observed that hatchlings could grow to some extent, but not optimally, in 100 percent seawater if they were offered fresh water every two weeks.

With these studies and also observation of hatchlings in the field, it is possible to propose a possible mechanism for the survival and growth of hatchlings during their first one or two years in a saline environment. The combination of a fully functional salt gland and the periodic access to fresh water, via rainfall, may be important in the overall viability of hatchling terrapins. A behavioral mechanism may also be at work: Hatchlings may not spend much time in the water. There is mounting evidence that the smallest terrapins spend most of their time hiding in the salt-marsh grasses or buried in the mud and are often out of the water except when flood tides inundate the marsh.

Although the mechanisms that allow diamondback terrapins to inhabit brackish environments are complex, they have provided this turtle with a unique advantage in comparison to other turtles. The diamondback terrapin has the marsh all to itself. It does not have to compete with other turtles or even with other reptiles for the nutrients and resources in the coastal habitats where it is found today.

## Temperature, Behavior, and Activity Cycles

In ectothermic reptiles such as terrapins, daily, seasonal, and geographically related temperatures have a profound effect on activity. Ectotherms cannot internally adjust body temperature. Their body temperatures are dictated by the environment. Terrapins do not have fatty insulation, fur, or feathers to help maintain heat, so they are also poikilothermic. This means that they cannot maintain a set body temperature via metabolic activity; their body temperature fluctuates and they can quickly gain or lose heat. Temperature is regulated more by behavioral, rather than physiological mechanisms. For aquatic turtles such as terrapins, basking is an important mechanism to increase body temperature. Terrapins are considered to be heliothermic; that is, they use solar heat to achieve optimum temperatures for physiological processes. Terrapins are sometimes seen basking on land, especially in the early spring, when water temperatures are still cool, but more often they are observed floating on the surface of the water, limbs splayed to expose as much skin as possible to the solar rays. From the terrapin's perspective, water temperature is more stable than air temperature. Basking in water allows body temperature to increase, even on windy days when basking on land can limit increases in body temperature and cause dehydration. Basking also allows terrapins to synthesize vitamin D, which is needed for metabolic regulation of calcium and phosphorus metabolism and thus for bone and shell health. Deficiency of vitamin D can cause soft shell disease, a condition sometimes seen when turtles are maintained in captivity without an ultraviolet light source. Basking is also important in decreasing the incidence of shell parasites such as algae and fungi, which can penetrate between the scutes and cause shell damage. Basking also facilitates the shedding of old keratin when growth of the shell occurs. As the keratinized scute material dries out, it peels back from the carapace and is readily sloughed from the shell (fig. 1.11).

The annual activity cycle of terrapins is dictated by water temperature and may be divided into several phases: spring emergence and breeding, summer dispersal, fall retrenchment into smaller creeks, and winter dormancy. In the north, warming of waters to approximately 13°C (55°F) awakens terrapins from their winter slumber. The timing and extent of terrapin activity phases vary somewhat among the subspecies. On Cape Cod, the northernmost subspecies has an extensive dormancy period, with almost half the year, late October to late April, spent in hibernation. While exploring Cape Cod's frozen Pleasant Bay during the winter, Elizabeth Hogan, photographer and naturalist, discovered a terrapin hibernation aberration. Contrary to the idea that Cape Cod terrapins are soundly locked into hibernation during the winter, Hogan found active terrapins while peering through a naturally formed hole in the ice. Apparently, an aquifer seep or a relatively warm spring was feeding fresh water into a section of the bay and allowing terrapins to maintain some winter activity in a very limited area.

A little farther south, on Cape May, New Jersey, terrapins are active for slightly longer periods. The winter habits of terrapins in the salt marshes of Cape May were studied by Yearicks, Wood, and Johnson (1981). By mid to late November, water drops from summer temperatures of 22 to 24°C (72 to 75°F) to 6 to 10°C (43 to 50°F). By late December, all signs of terrapins disappear from the larger bodies of water and the population hibernates in small creeks, 2 to 5 meters (approximately 6 to 16 ft.) wide. The terrapins do not emerge until April or May, and are not active even on relatively warm winter days. To find out where and how the terrapins were spending the winter, creek bottoms and creek banks were probed with rods and hard objects were excavated. In this manner, 311 terrapins were unearthed. All were alive. Cape May terrapins utilized different hibernating locations. Some were found in natural depressions on the bottom of creeks where they were covered by a thin layer of mud and 1.5 to 2.5 meters (5 to 8 feet) of water at low tide. Others buried themselves 0.15 to 0.5 meters (0.5 to 1.6 feet) deep into the sides of creek banks in areas free of vegetation and underground roots. Beneath undercut banks in the intertidal zone, group burials were found. In the latter cases, hibernating clusters were always covered by a thin layer of mud.

In Beaufort, North Carolina, hibernation was observed beginning October 22 and ending on March 8. A little farther north, in Crisfield, Maryland, terrapins emerged from hibernation on April 1 (Coker, 1906). The terrapins in these locations were occasionally observed to move in and out of hibernation, depending on the temperature, until mid-December when they remained dormant until spring. Although mating may occur within a restricted timeframe, Florida terrapins, especially the mangrove terrapins that inhabit the Keys, may be active year-round (Wood, 1992).

Terrapins are found at latitudes from about 26°N to 40°N. In the northern latitudes, they inhabit waters with extreme seasonal temperature variation. In contrast, relatively little temperature variation occurs in southern latitudes. From New Jersey to Massachusetts, surface waters may be frozen for extended periods (plate 4).

Hibernation, or brumation as it is sometimes called with reference to reptiles, is a type of dormancy exhibited by many turtles in temperate climates. Turtle hibernation is not a true hibernation, typical of endotherms. In winter, when food sources become scarce and growth and reproduction are not possible, endotherms hibernate to conserve energy. In preparation for hibernation, these animals often store fat. When they hibernate, physiological adjustments in their metabolism compensate for lack of food input. This mechanism lowers body temperatures and maintains the lowered temperature, producing a seasonal state of torpor, the inactive state associated with dormancy.

In contrast to true hibernation, the physiological mechanisms responsible for hibernation in turtles are less clear. It is not known whether they anticipate or respond to declining temperature by storing fat. Declining water temperatures rather than a shortage of food may be a more significant trigger for entering dormancy. Because turtles cannot maintain body temperature, they must avoid freezing, which can damage tissues.

For terrapins, hibernation is the response to plummeting temperatures. Almost all metabolic activity comes to a halt. At the onset of hibernation, terrapins take a breath of air and then initiate their last dive of the season. They then become dormant to survive periods of physiological challenge presented by decreasing water temperatures. There is no danger of starvation because even when active, terrapins can survive for months without food. Moreover, there is much less food available as water temperatures decrease. Under the muddy layers in tidal creeks, terrapins protect themselves from freezing. It was once observed that, for short periods of time, terrapins can withstand "cold severe enough to leave them encased in ice" (Coker, 1920). Somehow, they do not suffocate. Their need for oxygen declines as their metabolism plummets. Although there might be enough oxygen dissolved in the cold water to support the terrapin during this period of torpor, it is not clear how they survive in the anoxic mud into which they burrow.

During hibernation, physiological activity declines and the diamondback terrapin is not able to use behavioral mechanisms to osmoregulate. How does the ability of the terrapin to regulate salt balance and osmotic pressure change during hibernation? Gilles-Baillen (1973a) studied terrapins before, during, and after hibernation in either freshwater or seawater, and measured their osmotic pressure. As in other studies, higher osmotic pressures were measured throughout the year for seawater-maintained terrapins. The osmotic pressure of seawater terrapins increased throughout hibernation and did not decrease until after mid-April emergence from dormancy. The highest osmotic pressure correlated with passive entry of salt water at the time of emergence. The transient influx of sodium, seen during emergence and most likely due to incidental drinking, may actually serve an important function in reactivating the salt gland, which does not appear to function during hibernation. It seems that when terrapins are active, physiological mechanisms can maintain osmotic pressure of body fluids at a constant level, but during hibernation these mechanisms may be less effective. Nonetheless, hibernation has a successful outcome whether terrapins remain dormant in fresh water or salt water.

Dormancy in hibernating reptiles occurs in four stages. These stages, as they apply to diamondback terrapins, can be described as follows:

1. Fasting: Terrapins do not eat when temperatures dip below 15°C (59°F) (Davenport and Ward, 1993). When Yearicks et al. (1981) examined the gastrointestinal (GI) system of a subset of hibernating terrapins, they found their GI tracts to be empty. This observation agrees with findings from other hibernating turtle species. There is speculation that if food

were retained in the gut, it might lead to bacterial growth that could possibly produce internal damage.

2. Retreat to refugia or hibernacula: It is commonly observed that terrapins disappear from open waters as the temperature drops. They make their way into smaller creeks and burrow under the mud, where there will be less fluctuation in temperature than in the surrounding water.

3. Attainment of the dormant state: For northern diamondback terrapins, this may be a prominent feature of their annual activity (or lack of activity) cycle.

4. Metabolic depression: As food and air become inaccessible and as underwater temperatures further decline, metabolism slows down to conserve energy. Some dissolved oxygen can be obtained by gaseous exchange in the cloaca. If metabolism occurs without oxygen (anaerobic metabolism), lactic acid may accumulate and can be harmful if not eliminated. The extent of anaerobic metabolism that occurs in dormant terrapins is not known.

Turtles display another type of dormancy, called estivation, which serves as a behavioral response to dry heat. Terrapins in some areas will undergo estivation to survive heat and drought. The upper limit or critical thermal maximum for turtles is thought to be 41°C (105.8°F) (Alderton, 1988). During very hot weather, turtles will burrow, become dormant and exist in a state of torpor until the temperature cools enough for the turtles to emerge. In the Florida Keys, terrapins burrow in the marl (the clay and limestone muck typical of mangrove islands) in the dry season. For terrapins that populate other latitudes, the thermal stability of their aquatic habitat usually prevents large swings in environmental temperature within each season.

## Feeding and Nutrition

Terrapins are carnivores. In captivity, they are usually fed commercial turtle food, but their gusto for fish, crustaceans, and mollusks, if offered, is readily observed. Terrapins can be voracious eaters. When they are fed a satiation meal, they may eat up to 7 percent of their body weight but are ready to eat again after six hours (Davenport and Ward, 1993). There are several ways to study the dietary preferences of terrapins. Cafeteria-type feeding experiments can be conducted under laboratory conditions. Terrapins are offered several food choices and their preferences are observed. In the field, observers can take note of the food choices of feeding terrapins. If dead terrapins are found, dissections of their digestive tract may reveal a recent meal. Stomach flushing of live animals has been employed to identify dietary items. Lastly, fecal analysis can sometimes reveal the less digestible remains of terrapin prey. If captured and held for a short period of time (usually 24 hours is sufficient), terrapins will "donate" feces, commonly referred to as scat, to the researcher. Some investigators have described the construction of a device to sieve dried terrapin feces so that contents can be separated on the basis of size and identified (Bauer and Sadlier, 1992; Tucker and FitzSimmons, 1992). Using a combination of these methods for dietary analysis, we have some idea about the food preferences of terrapins.

A listing of the diet of terrapins reads like the menu of a seafood restaurant: crabs, snails, shrimp, fish, mussels, clams, and perhaps oysters. This list can be extended to include worms, insects, and carrion (Pope, 1946; Carr, 1952; Cook, 1989). Gastrointestinal dissection, stomach pumping, and fecal sampling sometimes reveal barnacles, algae, pieces of grass, and mud. The latter items are most likely incidentally swallowed with the more delectable food items.

Terrapins have no teeth; their strong horny jaws act as "seizers and choppers" (Pope, 1946). Terrapins have only been observed to eat under water. Claws help to tear the food apart, and muscular tongues manipulate food to the back of the mouth. Digestion is slow and temperature dependent.

In Beaufort, North Carolina, stomach contents of terrapins captured from their natural environment revealed that the major food item was the gastropod, *Littorina irrorata*, the periwinkle snail. In *Time of the Turtle*, Jack Rudloe's colorful account of encounters with turtles while collecting marine specimens in Florida, an elderly terrapin fisherman described his technique for finding terrapins in the thick marsh grass by "listening for the loud pops that came when they shattered a periwinkle snail between their powerful jaws" (Rudloe, 1979). The remains of *Melampus lineatus* (saltmarsh snail), small crabs such as the fiddler crab (*Uca*) and bits and pieces of annelid worms were also found in terrapin stomachs from Beaufort (Coker, 1906). A somewhat different diet may be typical of terrapins that live in northwestern Florida. Fecal analysis of 46 samples revealed a preference for dwarf surf clams (*Mulina lateralis*), crabs, and periwinkles, with crabs as the most frequently occurring food item, while surf clams were the major dietary constituent, comprising 83.1 percent of total fecal mass (Butler, 2000). In some areas of Chesapeake Bay, the main food items are soft-shelled clams (*Mya arenaria*), razor clams (*Tagelus* spp.), and smaller clams such as *Macoma* and *Gemma* (Roosenburg, 1994, Roosenburg et al., 1999).

Terrapins are not "sit-and-wait" feeders. After all, it is not often that a tasty periwinkle will saunter past a hungry terrapin. Eating only occurs in the water. High tides are thus prime meal times, contributing to the terrapin's daily activity cycle. Terrapins will be most active when the marsh is flooded by the tide and they can have access to submerged food sources. Bels, Davenport, and Renous (1998) studied how terrapins behave as mobile predators, stalking fast-moving and elusive prey such as shrimp, crabs, and fish. They described these activities as "strikes" and documented the modification of normal swimming activity when a strike occurs. Instead of using alternate limb strokes, like a dog paddle, the terrapin will transition to simultaneous action of the forelimbs, similar to the breast stroke, to provide a propulsive force. The change in stroke is accompanied by a rapid and large neck extension to decrease the distance to the swimming prey. The researchers describe the terrapin as a "ram feeder," rather than a suction feeder. It opens it jaws and overtakes its prey, rather than siphoning the food into its mouth like some of the jellyfish-eating sea turtles.

Terrapins eat crabs (*Uca, Callinectus, Carcinus*) but crabs can be dangerous prey. Davenport et al. (1992) observed that terrapins first make a visual assessment of the crabs: They are evaluating them for size. They then approach crabs with a maximum gape. If the crab is large, the terrapin will endeavor to crop off a limb and beat a hasty retreat.

In terms of less mobile prey, Tucker, Yeomans, and Gibbons (1997) wondered why terrapins prefer *Littorina irrorata*, which clings to the *Spartina* in the upper intertidal zone, when *Ilyanassa obsoleta*, the common mud snail, is so abundant and more accessible. What determines the preference for periwinkle in South Carolina marshes, where it can constitute 76 to 79 percent of the terrapin diet? It does not seem to be an issue of gape size. The answer seems to be the strength of the gastropod shell. When the compressive force required to crush the shells of the gastropods is compared, it becomes apparent that it requires 2 to 3 times more force to crush *Ilyanassa*. The researchers concluded that even though the energy cost in searching for *Ilyanassa* is less, the processing cost to consume the mud snail may deter terrapins.

It seems logical to predict that terrapins will cluster in areas with high food density, but this is not always the case. In a study of prey availability in several different creeks within a Connecticut salt marsh, Whitelaw and Zajac (2002) found that terrapin distribution in the marsh did not correlate with resource availability. It seems that other habitat factors such as tidal amplitudes of creeks and plant density may affect the distribution of terrapins within marsh systems.

There are no studies relating learned behaviors to feeding in adult terrapins but anecdotal reports point to a link. As mentioned previously, the sound of food being chopped up alerted farmed terrapins to a forthcoming meal. In addition to sound, other food-related stimuli may affect behavior. On some occasions when the feeding schedule of farm-raised terrapins was switched to evening hours, the appearance of a light source such as a flashlight brought terrapins "out in full force" (Coker, 1920).

Swiftly swimming fish are more difficult to catch than snails, but terrapins will catch fish if the opportunity arises. They have been observed feeding on Atlantic silversides (*Menidia menidia*) during spawning runs in the North Edisto River estuary in South Carolina (Middaugh, 1981).

Sexual dimorphism has an impact on the foraging ecology of adult diamondback terrapins. Because adult females are larger than juvenile females and males, it might be expected that they consume larger prey. In an effort to examine dietary preferences among different size classes of terrapins, Tucker, FitzSimmons, and Gibbons (1995) analyzed fecal samples from terrapins in the Kiawah Island marsh system in South Carolina. Data from captured terrapins were grouped according to gender, size, and age of the turtles. To achieve consistency, age was estimated by counting annuli on the right humeral scute of the plastron. Evidence for prey species was analyzed for occurrence, percent mass, and, in the case of *Littorina*, the number of opercula (covering of the shell opening) per terrapin. Undigested opercula, recovered from fecal samples, were used to calculate overall snail size. As expected, larger terrapins consumed larger snails. When consumption of *Littorina* was examined in female terrapins, size selection was apparent. Head size, and thus gape size and jaw strength, directly correlated with the size of the snails consumed. Small crab species, such as *U. pugnax* and *S. reticulatum*, were consumed by all terrapins, but only the medium and large terrapins ingested blue crab (*C. sapidus*), primarily by cropping their rear legs. Blue crabs were only a minor part of the total diet of the Kiawah Island terrapins. In this study, dietary diversity was found to be slightly higher in the largest size class of terrapins. Therefore, in female terrapins, resource acquisition can be expected to be related to body size. Small females consume fewer types of prey than larger females. The largest females are dietary generalists, consuming a wide variety of prey species. The fecal sieving technique does not allow researchers to accurately determine how soft-bodied prey, such as annelid worms, are represented in the diet. Since they did not find evidence of remains of annelids, such as indigestible setae or mandibular cuticles, Tucker et al. (1995) concluded that during their study period soft-bodied invertebrates were not a significant component of the terrapin diet. The researchers encourage ecologists to take more notice of the role of diamondback terrapins as macroconsumers in the salt-marsh food web. In marshes where terrapins are numerous, they may have significant impacts, particularly on *Littorina*.

## Growth, Development, and Life Span

Diamondback terrapins inhabit a large swath, approximately 1300 miles, of north to south coastal areas that experience dramatic seasonal differences in temperature. Therefore, it is not surprising to learn that the growth rates and age at maturity for terrapins vary with the subpopulation. The terrapins that have longer annual activity periods are expected to grow faster and mature at an earlier age. However, differences in growth rates and onset of maturity may even be seen within local populations.

It has been virtually impossible to assess the growth rates of wild terrapins during the first year or two after hatching. A laboratory study by Roosenburg and Kelley (1996), using hatchlings from eggs incubated at constant temperatures, points to increased growth rates in female hatchlings after a few months. Growth in both sexes will slow down as terrapins reach sexual maturity but this will occur in males at a much younger age than in females. After maturity, growth will occur at a low rate, less than 5 percent per year.

There have been several studies of growth rates in terrapins from different local populations. When Cagle (1952) plotted the growth rate of Louisiana terrapins and compared growth to the North Carolina population studied by Hildebrand (1932), he found that the growth rate was similar for the first two years, but after that, Louisiana terrapins grew faster than the North Carolina animals. In studying a population of Florida East Coast terrapins at the Merritt Island National Wildlife Refuge, Siegel (1984) measured greater growth rates than those of either the Louisiana or North Carolina terrapins. It should be noted that the North Carolina population consisted of captive animals that were fed by caretakers. Taken together, these trends would suggest that Northern terrapins grow more slowly than Southern subspecies, commensurate with the shorter growing season for Northern turtles. Other potential contributions to growth rate differences, such as local feeding habits, food availability, food quantity, and hatchling size, have not been fully compared across all the subspecies. Paradoxically, Northern female terrapins may grow more slowly than Southern ones, but they may achieve larger adult sizes.

Terrapins are assessed as "mature" by several criteria. For males and females, the appearance of secondary sexual characteristics is the hallmark of sexual maturity. When examining females, eggs can be detected by x-ray or palpation of the inguinal pocket. Nesting activity by females is also an indicator of sexual maturity. In young animals that have been sacrificed or found dead, one can dissect the specimens and perform histological staining to observe seminiferous tubules (vas deferens) in males and oocytes in females.

In looking at the north to south gradient, a noteworthy trend emerges with respect to maturation (table 1.2). In northern populations, the average size at maturity for females is 16.4 centimeters (6.5 inches) PL and 1063 grams (2.3 pounds), while their southern female cousins are slightly shorter and may weigh considerably less: 705 grams (1.6 pounds) for South Carolina females, 886 grams (2 pounds) for Florida Indian River females (Siegel, 1984). These North/South differences are less pronounced or may not exist for males. The increased size of some Northern females may have an impact on clutch size or egg size, an aspect of reproductive strategy that is explored in chapter 3.

The trend is for earlier age at maturation in terrapins as we move south along the eastern seaboard (table 1.2). From the Cape Cod, Massachusetts, subpopulation, south to Chesapeake Bay in Maryland, a female *M.t.terrapin* may be almost a decade old before she matures, while her precocious Southern cousins may mature in as little as four years. Northern males mature when they are five to eight years old, while Southern terrapin males are contributing to the gene pool when they are three years old. Rather than age, the primary determinant of sexual maturity is size, as can been seen when mean plastron length is compared. Females usually achieve about 14 centimeters (5.5 inches) PL; males must reach 8 to 9 centimeters (3 to 3.5 inches) PL before maturation is complete.

It appears that terrapin growth and maturation conform to what is seen in other aquatic turtles. The common trends are that: (1) Males mature earlier and at smaller sizes than females. (2) Growth is more rapid before maturity. (3) In temperate regions, Southern populations mature before Northern populations. (4) Sexual maturity relates more to size than age (summarized in Siegel, 1984).

Scientists and naturalists have always wondered about the longevity of turtles. These slow-growing, late-maturing creatures live long and potentially productive lives. There have been many stories, anecdotes, and even some documentation of terrestrial turtles that can be considered ancient by human standards. Aquatic turtles do not live as long as their land-dwelling relatives, but mark–recapture data have documented life spans of at least twenty years (Siegel, 1984; Wellfleet Bay Wildlife Sanctuary, unpublished records) for diamondback terrapins, and some researchers believe that they may live more than forty years. In populations that were studied in the early 1980s, marked individuals who were already mature at the time of capture can still be found, none the worse for wear. They do not seem to suffer from anatomical or physiological senility. Longevity may compensate for large variations in the success of reproduction from year to year (Gibbons, 1987). Certainly, older females continue to lay eggs and most likely do so throughout their life span. Since there is very high hatchling mortality in most species of turtles, it is the older, mature females that are lynchpins for the survival of the population. Unfortunately, these were the very specimens that were commercially valuable during the heyday of the commercial terrapin fishing industry.

Although much is known about the diamondback terrapin, there are considerable gaps in our understanding of how this turtle copes with stressful environmental conditions. The many studies on osmoregulation point to a suite of physiological and behavioral adaptations, but it is not clear how these mechanisms work in concert. The physiological aspects of hibernation and estivation are partially understood, and much of what we know is based on work with other turtles. The significance of the size differential between terrapins of the various subspecies is without explanation. Further research may help to answer these aspects of terrapin anatomy, physiology, and growth.

# Chapter 2

## A Coast-Hugging Turtle

THE DIAMONDBACK TERRAPIN inhabits brackish waters bordering prime real estate along a narrow ribbon of coastline that traces the Atlantic Ocean and Gulf of Mexico. The range extends from north temperate to subtropical climate zones and is generally classified as salt marsh in most locations. The most extreme variation of the habitat can be found in the mangrove swamps of Florida Bay and the Florida Keys, which are home to the southernmost population of terrapins. Despite the dense human population in coastal communities, very few coastal inhabitants and visitors have actually seen diamondback terrapins in their natural setting. A closer look at the salt marsh and mangrove swamp reveals why these are ideal terrapin locales and clarifies the importance of the health of these regions for the survival of *Malaclemys terrapin*.

### The Atlantic Salt Marsh

The smell of low tide in a salt marsh is so distinctive that those who live near one can often tell the status of the tides by using only olfaction. If it's not too powerful, I actually enjoy this marshy smell, perhaps because I associate it with summers near the beach, but some folks, justifiably, find it offensive. The marshy odor, sometimes very strong and most noticeable at low tide when the flats are exposed, is due to the production of hydrogen sulfide ($H_2S$) by bacteria that reside in the darkly colored sediments just below the surface. Marsh sediments are composed of minute particles, so small and so compacted that very little oxygen is present. The bacteria that have adapted to this hostile environment have figured out how to use sulfate for their metabolism similar to the way we use oxygen, but the resulting end product is smelly $H_2S$ rather than odorless water, $H_2O$. Hydrogen sulfide is a colorless gas that easily wafts into the air and disperses all around the marsh and neighboring areas. It is the same gas produced during the decomposition of eggs.

Salt marshes form where there are indentations along the coastline, often in regions that are partially surrounded by land and sheltered from wave action. These locations include areas protected by sand bars and barrier islands as well as the land bordering bays, inlets, and coves. The daily cycle of the marsh is dictated by the semidiurnal tides; the alternation of high tides with low tides occurs twice a day at intervals of approximately six hours. The magnitude of tidal variation is determined by location. In New England and Georgia, the difference between the mean high and low water marks averages about 3 meters (almost 10 feet), while in the mid-Atlantic states and Florida, tides may vary less than 1 to 2 meters (3.2 to 6.5 feet). During low tide, the area of the marsh that was under water at high tide becomes exposed. This area washed by the tide is referred to as the intertidal zone and contains a unique collection of organisms that have adjusted to and may actually prefer periodic flooding. The tides have a major effect on the salinity of the marsh. During low tide, on a hot, sunny, summer day, the salinity of the marsh may increase to 60 ppt, twice that of seawater, while after heavy rains, the salinity may decrease to less than half that of seawater.

Some salt marshes are located near estuaries where fresh flowing river water flows into the ocean or bay. The geographic boundary of the estuary is usually drawn upriver where the salinity decreases to 0.5 ppt, a condition that is considered to be fresh (Berrill and Berrill, 1981). Subject to tidal flow, the estuary is constantly remolded by sediments, fluctuating water levels, and the constant mixing of fresh and salt water. In addition to changeable salinity, estuarine water temperatures are quite variable. Unstable salinity and temperature are not tolerated by many organisms. However, the nutrient-rich estuary can promote the growth of tolerant species of phytoplankton that are part of an important food web. The phytoplankton serve as food for larvae of invertebrates, including worms, snails, mollusks, and crustaceans, that in turn become food for fish and diamondback terrapins.

On visual inspection, the salt marsh is dominated by rooted, salt-tolerant plants in the genus *Spartina* (plate 5). One type of Spartina is low-growing salt-marsh hay, *Spartina patens*. This grass may only be fully exposed to salt water twice a month, during high spring tides. These brief periods of salt exposure are critical because they allow *S. patens* to retain its foothold in the marsh and not be overtaken by freshwater marsh plants whose roots cannot tolerate even small amounts of salt. Expanses of the short, thin blades of this dominant salt-marsh plant form luxuriant meadows. We often see masses of *Spartina patens* flattened down by the wind in whirled patterns known as cowlicks because they resemble similar patterns that can characterize human hair (Teal and Teal, 1969). In the 1630s, salt hay was the resource that attracted early Cape Cod settlers. Although the Pilgrims initially abandoned the Cape after the Mayflower first landed in Provincetown, they remembered the marshes, and eventually returned to use the meadowlands as pasture or to cut and collect hay to feed their livestock.

*Spartina alterniflora*, often referred to as cordgrass, prefers a wetter substrate. It grows in tall, stiff, dark green spikes and tends to be partially covered during each high tide. Its stems, 1 to 2 meters (3.3 to 6.5 feet) tall, were used by colonists as thatch for roofs. Sprinkled among the *Spartina* are other, often solitary salt-marsh plants, most of which are found in the higher, drier areas of the marsh. These include spikegrass (*Distichlis*), black needle rush (*Juncus*), the woody stemmed marsh elder (*Iva*), the leafless succulent glasswort (*Salicornia*), the spear-leaved arrow plant (*Atriplex*), and the late summer-blooming plants such as sea lavender (*Limonium*), prized for dried floral arrangements, the marsh aster (*Aster*), and seaside goldenrod (*Solidago*). On the seaward side of the marsh, eelgrass (*Zostera marina*) beds help to trap sediments, allow the marsh to grow, and serve as a nursery for many aquatic organisms.

With their expanses of plants and algae, salt marshes are very productive habitats. *Spartina* is an important component of the marsh food web. In the absence of large herbivores to graze on stems and leaves, *Spartina* becomes food indirectly, after its outer leaves die. The invisible single-celled, microscopic marsh organisms, consisting of many types of bacteria, transform the grasses to detritus. Coupled with algae, the detritus completes the meal for countless invertebrates such as worms, crabs and mollusks. These, in turn, are consumed by carnivores such as fish and diamondback terrapins.

In the North, the marsh vista varies during different seasons of the year. Plants in North Atlantic marshes have a fast growth rate throughout the spring and summer months but then experience complete dieback during the winter. At the northernmost range of the diamondback terrapin, on Cape Cod, ice often covers the marshes for at least part of the winter (plate 4). The marsh substrate may freeze to depths of 20 to 25 centimeters (about 8 to 10 inches) (Berrill and Berrill, 1981). As the ice breaks up in spring, some of the marsh surface can break up with it. Considerable amounts of marsh can be lost during these "ice scouring" events. A walk through the Cape Cod marsh during the spring thaw can be like walking through a surreal landscape of dead, brown grasses interspersed with large mounds of recently buckled mud thrown up from shallow creek bottoms. One wonders if any terrapins have been disturbed from their hibernation during this upheaval. Occasionally, I have found terrapin carcasses on the marsh in early spring, soon after the thaw, possible victims of frost heaves or unsuccessful hibernation. In Southeastern marshes, plant . dieback is not as dramatic. The older, browner grasses are slowly replaced by the newer, greener growth that sprouts up even in winter. In the mangrove swamps of Florida, the difference between summer and winter is not visibly remarkable but can certainly be distinguished by the creatures that respond to seasonal changes in terms of mating and other behaviors.

Hidden among the plants are the animals that are part of the salt-marsh community and are necessary for the health of the marsh. Some of these animals are also prey for diamondback terrapins. The salt- and temperature-tolerant animal community of the salt marsh and estuary include burrowers, grazers, and foragers. Embedded in the mud are polychaete worms, recognized as a good source of bait by fishermen, and filter feeders such as ribbed mussels (*Geukensia demissus*) and various types of clams. Ribbed mussels are not considered edible by humans but are eaten by Cape Cod terrapins. We find fragments of ribbed mussel shells in adult terrapin fecal samples. Our tiny, lab-raised hatchlings will devour the soft parts of ribbed mussels if we help them out by first cracking the mussels apart. Another filter feeder, the American oyster (*Crassostrea virginia*), is found atop the flats, providing there is substrate for attachment.

Although there have not been many studies done to identify cordgrass predators, some snails and insects feed on cordgrass, sucking sap and eating leaves. *Melampus bidentatus*, the tiny salt-marsh snail, breathes air via lungs. Sensitive to drying out, it is sometimes found under marsh vegetation, where it can offer up a meal to a terrapin hatchling. As the tide rises, *Melampus* inches up the blades of *Spartina* to keep one step ahead of the rising water. As the tide ebbs, *Melampus* slowly slithers down the Spartina blades, remaining just above the water line. While transitioning along the blades, *Melampus* is sometimes just within reach of terrapins that are feeding near the marsh surface. Periwinkle snails, belonging to the genus *Littorina*, also travel up and down *Spartina* stalks with the changing tides and are common marsh grazers. Their tendency to travel up the stalks when tides rise helps to keep some of them out of reach of hungry crabs and terrapins. *Littorina littorea* is found in New England marshes; it is replaced by *Littorina irrorata* in the South and *Littorina angulifera* in Florida Bay. The latter species crawls up and down the aerial mangrove roots. *Littorina* provides food for adult diamondback terrapins whose strong jaws can crush the shells. In some locations, such as Cape Cod (Brennessel et al., 2004) and Long Island (Draud, personal communication), hatchlings and juveniles rely on the smaller *Melampus* for nutritional needs.

In some cases, diamondback terrapins can be categorized as secondary consumers in the marsh food web. This means that terrapins eat the animals that eat the plants and detritus. Removal of terrapins from a salt marsh would be expected to have an impact on the producers (plants and algae) and the primary consumers (predominantly snails). Without terrapins, the number of snails may increase. This could lead to unchecked grazing of plants, the decline of the salt-marsh community, and eventually the decline of the marsh itself. Terrapins are also considered to be tertiary consumers because they eat crabs and fish. Thus, terrapins have a complex role in the marsh food web.

Salt-marsh foragers such as crabs are prey for terrapins. Various crabs live in burrows and feed on worms, mollusks and smaller crustaceans. In marshes where diamondback terrapins are found, the common crab species include the invasive green crab, *Carcinus maenus*, the highly sought blue crab, *Callinectus sapidus*, and most abundantly, fiddler crabs of the genus *Uca*. Fecal samples of even the smallest terrapins can reveal crab parts. Juvenile terrapins eat the smaller crabs, while larger terrapins are known to tackle adult blue crabs by using their strong jaws for selective limb cropping. Most marsh visitors have seen fiddler crabs and recognize the asymmetric male, who has one very large claw that he waves aggressively to defend his burrow and attract females. If he is lucky, a female will follow him into his burrow to mate. Burrowing animals such as fiddler crabs have been shown to stimulate growth of cordgrass, most likely by improving drainage and aeration of the marsh substrate.

Those of us who spend time around salt marshes become well acquainted with its resident insects. Some of the marsh insects introduce themselves to us in a very unpleasant manner. Perhaps the most annoying insect, from a human point of view, is the salt-marsh mosquito. *Aedes sollicitans*, the Eastern salt-marsh mosquito, inflicts pain and itchy discomfort on visitors from Cape Cod to Texas, while *Aedes taeniorhynchus*, the black salt-marsh mosquito, is more predominant in the southern United States, from North Carolina to Florida. Although the male mosquito is harmless to humans, the females are so troublesome that beginning in the 1700s, many salt marshes were ditched and drained. These hydrologic alterations are discussed in chapter 5. The rationale was to eliminate breeding grounds for mosquitoes that cause malaria. Unfortunately, these attempts have sometimes backfired. With marshes becoming dryer, small fish that prey on mosquitoes are less able to survive. The female mosquito, however, can still breed in transient films on the surface of drained marshes.

Another marsh insect that can make life miserable for humans is the large greenhead fly. The males feed on *Spartina* juices, but the females look to animals and humans for a blood meal. For some reason, the greenhead is attracted to objects that are blue, so one is advised not to wear blue in a salt marsh in early summer. In some locations, this attraction to blue objects is used as a form of greenhead pest control. In several of the Wellfleet marshes, raised blue boxes dot the landscape. Unwary greenheads make their way into the boxes and are unable to navigate an exit, and thus perish. One of the smallest marsh insects is the midge. Midge is a generic name for fragile flies that actually belong to several different insect families. The genus *Chrirono-mus* contains organisms that are often seen in large swarms on the surface of the water. The biting midges of the genus *Culicoides* belong to a different insect family. These blood-sucking pests, commonly called punkies, sandflies, or no-see-ums, are responsible for some less-than-elegant choreography performed by marsh visitors as they move, jump, swat, scratch, and shake all parts of their body in response to the bite of the barely visible pest. It is so tiny that it can make its way through porch and window screens.

When I bring friends and colleagues to the marshes to look for nesting terrapins or to track hatchlings, I find that while I am always looking down, they are looking up, attracted to the most vocal and visible marsh creatures, that is, all the birds that reside in or visit the marsh in their travels. Residents and migrants alike make the marsh a vital, noisy place, with their flying, skittering, wading, pecking, scuttling, squawking, shrieking, and diving. So much has been written about marsh birds that I will not describe them here, but I would like to highlight a few associations between birds and diamondback terrapins. It is often the case on Cape Cod that when I find a terrapin nest on the marsh shore of a barrier beach or along the shore of a large creek, I find piping plovers nesting on the opposite side of the dune. I can stand at the top of the dune and see two threatened species at the same time! Although diamondback terrapins have few avian predators, the bald eagle may possibly be one of them. In Florida Bay, within the Everglades National Park, shells of smaller diamondback terrapins, most likely males and/or juveniles, are occasionally spotted in bald eagle nests. It is not known whether the terrapins served as eagle prey or if they were scavenged after another predator had first pickings. Crows and gulls sometimes make a meal of terrapin hatchlings.

Ubiquitous raccoons dig for crabs and clams and are predators of diamondback terrapin eggs. Various rodents, such as rice rats (in the South), black rats (in the mangroves), Norway rats (on Long Island, New York), and white-footed mice (in the North) also find a home in the marsh and may feast on terrapin eggs and hatchlings.

## Cape Cod

Except for one terrapin "spotting," on Boston's North Shore, diamondback terrapin populations are not known to occur north of Cape Cod, Massachusetts. Although Massachusetts also has Cape Ann, home to the famous Gloucester fishermen, when New Englanders refer to "the Cape," it can only mean Cape Cod, a thin stretch of land that resembles a flexed arm with Provincetown at its fingertip. Only a few miles wide in places, the Outer Cape was called the "Narrowland" by the original inhabitants, who hunted game, fished, and cultivated beans and corn. A cache of native corn, discovered in a shallow burial area in the Truro Hills, helped the Mayflower pilgrims survive when they first arrived in the new world.

Terrapins are not evenly dispersed throughout the Cape. Three main populations can be found. Pleasant Bay, Orleans, site of the largest embayment on the Cape, is home to a group of terrapins that stem from a population that was historically part of a modest terrapin fishery. Pleasant Bay terrapin spotters work mostly from kayaks; very little nesting activity has been observed. The area around Pleasant Bay has seen many changes over the years. The greatest potential impact from a terrapin's point may be the extensive development that has occurred around the bay. In addition, storms, such as the one that occurred in 1987, pound the area and create breaches in the barrier beach that protect the bay from an influx of ocean water. This major breach is responsible for current changes in tidal heights and probable loss of uplands in terrapin nesting areas. Just north of Pleasant Bay, terrapins were historically observed in Nauset Marsh but none have been seen recently. The last known hatchling from the area was photographed in 1976. More than likely, this population is now extirpated. Another cluster of terrapins inhabits the inner elbow of Cape Cod, and they can be found on the bayside marshes of Eastham, Orleans, and Brewster.

The town of Barnstable also had a small terrapin fishery that supplied turtles to restaurants in Boston and New York. The Barnstable marsh is located between the six-mile long barrier beach known as Sandy Neck, and the mainland. The terrapins of Sandy Neck have been extensively studied since the late 1970s by Peter Auger, who has used terrapins as a case study for his biology classes at Barnstable High School and for the field research courses that he teaches at Boston College. Based on long-term observations of terrapin nesting activity, Auger has estimated the population to consist of approximately 1,000 nesting females. Auger and his students have piloted a Cape Cod head-starting program to raise hatchlings over the winter. They are concerned about late-season nests, that is, second-clutch eggs that are laid late in July and may not develop until late fall. Auger believes that there is high mortality in these nests. When Auger and his students allow eggs to complete incubation in the laboratory, raise hatchlings over the winter and release them in spring, terrapins may have a better chance of surviving their first year. (The perils of young terrapins and the rationale for headstarting are explained in chapter 4.)

Wellfleet Harbor, about twenty nautical miles from Sandy Neck, is also home to terrapins. There are no historic records that point to a terrapin fishery in Wellfleet, although whaling, oystering, scalloping and other commercial fishing ventures have characterized Wellfleet since its establishment in 1763 when it officially separated from the town of Eastham. Today, Wellfleet is most famous for its art galleries, oysters, and its ocean beaches bordered by majestic dunes. Wellfleet Harbor, less than two miles from the ocean, across the Narrowland, is a large sheltered cove on Cape Cod Bay. It is bordered on the northwest by Great Island, which narrows out to Jeremy Point, an area that has been eroding at a rapid rate. Off the tip of Jeremy Point, the meager remnants of the settlement on Billingsgate Island can still be seen at low tide. A narrow stretch called the "Gut," technically known as a tombolo, connects Great Island to Griffin Island and completes the northern section of the harbor. This portion of Wellfleet is part of the Cape Cod National Seashore. The business area of Wellfleet, situated on the eastern side of the harbor, is a busy place during the summer months. Wellfleet is known to locals and vacationers alike as the "art gallery" town. The small pier and marina support a year-round fishing industry but gear up for recreational use during the summer. Sailboat and motorboat moorings, fishing charters, boat rentals, and a few restaurants characterize the surroundings. Heavily used by the local townspeople, the harbor is also the setting for town recreational and arts space: skate park, tennis courts, playgrounds, and birthplace of WHAT, Wellfleet Harbor Actors Theater. Wellfleet terrapins coexist with lots of people and lots of activities. Most of the time, the terrapins remain unnoticed as they go about their business. To the south of the marina, Indian Neck forms a trapezoid-shaped peninsula. The region is called Indian Neck with reference to the historic settling of the area by the natives as they were driven from more "desirable" areas by European settlers. Although a quiet part of Wellfleet, Indian Neck has experienced extensive development over the past fifty years. Across Blackfish Creek from Indian Neck, Lieutenant Island rises from the landscape. Reached by a small bridge, Lieutenant Island causeway becomes flooded at high tide. Residents must pay careful attention to the tide charts as they plan their comings and goings. There is a very extensive marsh on the south side of Lieutenant Island within which many smaller creeks and drainages feed into Cape Cod Bay. South of the harbor, stretching from Route 6 to Cape Cod Bay, we find the Wellfleet Bay Wildlife Sanctuary (WBWS), one node in the network of Massachusetts Audubon Sanctuaries. Having visited many Massachusetts Audubon Sanctuaries, I can truly say that this one is special. It has developed from an ornithological station into a mecca for birders, hikers and vacationers. It supports a wide variety of programs that appeal to nature lovers of all ages.

In summer, ocean temperatures rarely rise above 15.5°C (60°F) and surfers are wise to wear wet suits, while the shallows of the harbor can reach a toasty 27°C (80°F). Three-meter (about 10-feet) tides are common in Wellfleet. When the tide ebbs, acres and acres of marsh and mud are exposed. Unless boaters have had experience in the harbor or have been warned about the tides, they can very easily become stranded for hours on a sand bar, victims of a falling tide that rapidly disappeared beneath their boat. Some believe that it is the flushing action of these massive tides that makes Wellfleet a prime location for shellfish and causes Wellfleet oysters to be so sweet. Because the harbor is so shallow, some vessels cannot leave or enter the main docking area for one or two hours on either side of low tide. If undisturbed, the harbor would fill with sediment over time, so periodically the Army Corps of Engineers dredges to maintain navigation channels. The irregular coastline of Wellfleet Harbor is approximately twenty-five miles long. Here and there, mysterious-looking boxes can be seen on the flats at low tide. These are oyster trays, used by aquaculture entrepreneurs to raise valuable shellfish in an environment that protects them from some of their predators.

Working as a herpetologist for the Massachusetts Audubon Society, Dr. James "Skip" Lazell alerted researchers to the presence of terrapins in Wellfleet Bay. This population has been studied since 1980. Wellfleet terrapins emerge from hibernation in mid to late April, when the water temperature reaches 13 to 15°C (mid to high fifties Fahrenheit), but they may disappear for days at a time if the temperatures drop precipitously, as is characteristic of the fickle New England springtime. Even when the inland air temperatures are balmy, east winds whipping across the offshore Labrador current may keep Wellfleet air and water temperatures below a temperature that is comfortable for terrapins.

Led by Bob Prescott, scientists and naturalists from WBWS have dispatched volunteers and summer campers to search for terrapins. Over the years, a number of volunteers and interns have kept tabs on the nesting females. In the late 1990s, Don Lewis moved to Wellfleet and became a WBWS volunteer and directed some of his boundless energy to the study and protection of turtles. He is one of the leaders of the sanctuary's Sea Turtle Rescue program during the winter months, but he began to spearhead terrapin conservation and education efforts during the warmer months of the year. Lewis developed an informative web site (http://www.terrapindiary.org) and posts wonderfully written stories and glorious photos describing terrapin seasonal activity. Lewis has expressed the rationale behind his affinity for turtles:

*Fig. 2.1. Marginal scute markings can be used to identify this terrapin.*

"Because when you are retired and looking for a research target, you ought to pick a critter you have a reasonable chance of catching." In order to conduct a population study, Lewis pioneered the use of canoes and kayaks to net terrapins from the shallows. This is a much more challenging task than one can imagine. There is only a brief window of opportunity in spring and early summer when the water in Wellfleet Harbor is clear enough to see submerged terrapins. The slightest breeze can ripple the water enough to obscure visibility and if the tide is high, the terrapins take a deep dive and escape encroaching nets. During extremely low spring tides, Prescott and Lewis take visitors and volunteers to Blackfish Creek, named after the pilot whales (blackfish) that strand on Cape Cod Bay beaches and were used by natives and settlers for extraction of whale oil, a process known as "trying." This large creek extends from its source near Route 6 to Cape Cod Bay between Lieutenant Island and Indian Neck. Although a bit over 500 meters (about 0.33 miles) wide at high tide, the creek becomes a narrow funnel about 25 meters (approximately 80 feet) wide and a meter deep during a low spring tide. All the smaller drainages feeding into Blackfish Creek, such as Paine Creek, Drummer Cove, and Loagy Bay, become completely dry. One can wade into Blackfish Creek and net terrapins as they are flushed from smaller, shallow upstream creeks by the tide. Prescott and Lewis have also employed a seine net, stretched across the creek and held in place by strong volunteers, to catch terrapins during their trip into the deep-water mouth of the creek. Neither dip netting nor seine netting for terrapins is very efficient; many more escape than are captured, but the quest can be an enjoyable spring activity. If the low tide occurs early, we get treated to a beautiful sunrise as we look to the source of the creek; in the evening, the sun will set over the water where the creek meets Cape Cod Bay.

To get a sense of population numbers, gender ratios, and age distribution, each terrapin is "processed." The terrapin receives a unique number, and it is "marked" by notching the shell with a metal file. A variation of the marking system of Cagle (1939) has been adopted in which notches are made between or within marginal scutes (fig. 2.1). Over 1,500 Wellfleet terrapins have been thusly marked and can be identified when they are recaptured. The total population size is not known but probably numbers in the thousands, with a gender ratio of two to three females for each male. A caliper is used to measure the turtle's straight-line carapace and plastron lengths and its carapace width in two locations, the widest part and at the suture between the first and second costal scutes. These carapace measurements can give an index of how the shape of the carapace changes as a terrapin ages. The width of the plastron between the bridges is also measured. Each terrapin is weighed, structural anomalies are noted, and digital photos are taken. The photographs are very useful for year-to-year comparisons of overall growth, observing new injuries or healing of old injuries, and as an aid to identification if injury to shell has obscured the markings. Lewis's photographic memory sometimes makes the digital images redundant. He can usually remember each terrapin, where it was captured and the circumstances of the day. He also has an uncanny ability to recall the weights and measurements of some of the turtles.

Wellfleet terrapins can be found in four nesting clusters. The northern cluster nests within the Cape Cod National Seashore in an area that abuts the Herring River estuary. Terrapins nest on Great and Griffin islands and on the narrow "Gut" that links the two islands. Terrapins also nest on Indian Neck peninsula, with very concentrated nesting found in a few locations very close to the Blackfish Creek marshes. Lieutenant Island is heavily used by nesting terrapins, although certain locations are favored over others. Lastly, Wellfleet Bay Wildlife Sanctuary and adjacent marsh uplands in Eastham near First Encounter Beach, whose name commemorates the first Pilgrim–native meeting place, form another nesting cluster. Volunteers and interns are dispatched to walk the marsh paths and beaches in these locations from mid to late June until mid to late July to observe nesting females. Some have put in many hours on the terrapin team. Liz Moon has been a dedicated terrapin volunteer for many years, devoting her summer vacation weeks to many miles of walking and observing. Lewis and the WBWS have set up a terrapin hotline so that folks who spot terrapins can call in. Someone will be dispatched to confirm the sighting or occasionally to find that a snapping, box, or painted turtle was mistaken for a diamondback terrapin. In general, Wellfleet terrapins are homebodies. A few have been found at distances four to six miles (around eight to ten kilometers) from their original site of capture, but most are found in the same creeks and marshes from year to year. In late August, volunteers and interns begin nesting patrols to look for signs of hatchling emergence. Terrapin research and conservation in Wellfleet have been supported in recent years by Friends of Pleasant Bay, Wheaton College, the Sounds Conservancy, and the National Heritage and Endangered Species Program. There is a great sense of camaraderie among researchers, interns, and volunteers who study and try to protect the terrapins in Wellfleet, at the northern edge of the existence of the species.

Historical records (Babcock, 1926, 1938) place terrapins on Nantucket Island and also in Buzzards Bay. Although current surveys do not report terrapins on the islands off the coast of Cape Cod (Nantucket and Martha's Vineyard), recent collaborative efforts spearheaded by marine science teacher Sue Nourse of Tabor Academy, veterinarian Mike Ryer of the Buttonwood Park Zoo in New Bedford, and Mark Mello, director of the Lloyd Center for the Environment, have been successful in locating terrapins on the western shores of Buzzards Bay. There are reports that 100 diamondback terrapins from southern New Jersey were introduced into this area in 1972 by a well-meaning college professor (Lazell, 1976). It is not known whether the Buzzards Bay colony is a remnant of a historical grouping or the progeny of introduced turtles. Genetic studies have the potential to shed some light on their origin.

## Rhode Island and Connecticut

In the tiny state of Rhode Island, the terrapin population of Hundred Acre Cove was monitored throughout the 1990s. This large salt marsh in Barrington is believed to be the last remaining terrapin stronghold in the state. Efforts are underway to create a wildlife refuge that will be sufficient to sustain the terrapin population. Studied over an eleven-year period, the Barrington site has approximately 230 nests laid per year by about 188 breeding females. Although the breeding female population was found to be stable during the period of the study, recruitment of younger females into the adult population was found to be decreasing. The cause for the low recruitment could not be determined but it seems that low survival of eggs, hatchlings and juveniles is most likely an important contributor (Mitro, 2003).

Diamondback terrapins have been historically distributed along the Connecticut and New York borders of Long Island Sound and the area of New York State around the Hudson River estuary. There have been terrapin sightings sixty-nine kilometers (forty-three miles) up the Hudson River near Peekskill, New York (Klemens, 1993). Their presence was documented in the 1940s in various Connecticut locales and along the creeks and shores of the Connecticut River (Finneran, 1948). In the 1990s, terrapins could be found along the Connecticut coast in locations west of the Connecticut River, which bisects the state from north to south. The turtles were found in polluted waters, and they often clustered around the warm water discharge outputs of power stations along the Connecticut shoreline, where, according to the Connecticut Department of Environmental Protection, they congregated in large numbers. In areas east of the Connecticut River, localized populations have been reported (Klemens, 1993). Investigators at Fairfield University have studied diamondback terrapins in a complex system of marshes in Milford and Stratford where they report 100 percent site fidelity of terrapins to home creeks (Gauthier et al., 2000).

## New York

Terrapin remains have been found in excavations of prehistoric native villages at Orient Point, on the east end of the North Fork of Long Island, and on Shelter Island, between the North Fork and South Fork. The diamondback terrapin was an asset to the Long Island economy in the early 1900s, but after being aggressively harvested, it became quite scarce, and some believed that it was extirpated (locally extinct) by 1930. The finding of a terrapin egg in 1969 in South Oyster Bay provided the first hint that the species had not been completely eliminated (Spagnoli and Marganoff, 1975). Today, Long Island terrapins are found in small clusters on the North Fork and South Fork: Oyster Bay, Mount Sinai Harbor, South Oyster Bay near Fire Island, Peconic Bay near Riverhead, Shirley, Captree Basin, Huntington and Nesconset rivers, Cold Spring Harbor, and other locations.

Researchers have returned to areas of Oyster Bay and Mount Sinai Harbor, historic homes for terrapins on the north shore of Long Island, and are assessing the present populations. Similar to what is known to have occurred in Buzzards Bay, Massachusetts, there is a persistent rumor that terrapins were reintroduced to these areas in the early 1900s from stocks of farm-raised turtles. In the affluent town of Oyster Bay, luxurious estates are scattered on the low-lying hills across the water from the local town beach. Dr. Matt Draud of C. W. Post–Long Island University was our gracious host as we visited his study area. Draud and I are doing similar studies on tracking terrapin hatchlings, so I took my Wheaton College and Sounds Conservancy interns to visit his study area and talk shop. Energetic Draud has several terrapin projects running concurrently. He and his C. W. Post team followed up on some initial studies by graduate student Marc Bossert (Draud and Bossert, 2002). They conduct mark/recapture studies, perform sonic tracking, stay up to all hours of the night to observe nesting activities, and crawl around in the marsh on hands and knees looking for small turtles. Draud has an excellent handle on the Oyster Bay terrapin population. He knows where they nest, where they hibernate, how they move about, and how they are threatened. Draud took us to Centre Island Beach, where several terrapin nesting areas can be found. Draud and his former graduate student and Sounds Conservancy intern Barbara Bauer had been at the beach from 4 to 7 A.M. that morning, during high tide, and found several nesting females. The small patch of beach that is used for nesting in June is covered with gravel (plate 6) and abuts a parking lot where a carnival, complete with rides and amusements, was set up. It seems that the timing of the annual carnival coincides with terrapin nesting in Oyster Bay. The adjoining marsh is not extensive; the swath of *Spartina patens* is only a meter wide. Yet nesting terrapins are drawn to this small area of gravel and sand. Draud has photos of a terrapin nesting in close proximity to a family enjoying a day at the beach. There is a high rate of nesting success at this site. Draud speculates that it may be because the beach is a local nighttime hangout for teens. Perhaps the presence of the teens deters potential predators.

On an adjacent beach, owned by a local gun club, nests are depredated at a high rate. Draud sometimes relocates nests that he finds in this area to the town beach. If he finds female terrapins that have not yet laid their eggs, he will occasionally bring them to his laboratory at C. W. Post and induce the females with oxytocin. Once they lay their eggs, he buries them on the town beach in a location where they have a good chance of survival.

Although these nests produce hatchlings, Draud has identified an unusual problem for terrapins in Oyster Bay. The Norway rat (*Rattus norvegicus*) has cued into the presence of terrapin hatchlings in fall, and again in late spring as the hatchlings emerge from hibernation. Draud has documented a high level of hatchling mortality due to rats (Draud, Bossert, and Zimnavoda, 2004). Bauer described the feeling of being at a terrapin nesting site in the middle of the night and finding many pairs of beady eyes staring at her as she scanned the marsh looking for terrapins. She once mistook a swimming rat for a terrapin during an attempt to catch turtles by snorkeling in the offshore channel. Speaking as someone who does not have a fondness for rats (even though they are warm and fuzzy), I think the C. W. Post interns should receive hazard pay for their contributions to the research effort.

By using sonic tracking, Draud has followed terrapins into hibernation and has located a group hibernaculum in a small region of the harbor that remains under a few feet of water even during low tides. He has found the terrapins stacked on top of one another as they spend their winter in the harbor sediments. As we traveled by boat to the area of the harbor where Oyster Bay terrapins hibernate, Draud described plans by the town to dredge the harbor to create additional boat moorings. With the support of a local conservation group, Draud has asked the local authorities to schedule the dredging operations so that the hibernating terrapins are not adversely impacted.

The Oyster Bay population has equal numbers of males and females. For the past decade the terrapin population in Oyster Bay seems to be steady, but Draud's data indicate that the population may be aging. Similar to the status of the Rhode Island colony, this suggests that recruitment is low and that predators and other threats may be gaining the upper hand.

The island of Manhattan is the usual destination when tourists visit New York City. However, there is another island in the Big Apple that diamond-back terrapins visit once or twice each year to lay their eggs. Ruler's Bar Hassock is the largest of the tiny islands in Jamaica Bay and serves as nesting area for the most remarkable terrapin colony one is likely to find. Some of the other neighboring islands are covered during high tide and are considered to be marshland. Black Wall Marsh, Little Egg Marsh, The Raunt, JoCo Marsh, and Pumpkin Patch Marsh are some of the tiny islands that make up Jamaica Bay Wildlife Refuge (JBWR), 1,200 acres of preserved land in the southern part of Queens, New York. This refuge, part of Gateway National Recreation Area, is sandwiched between two airports: Floyd Bennett Field, which is a decommissioned Air Force Base, and bustling John F. Kennedy (JFK) International Airport. The sanctuary is accessible by public transportation. Visitors can take the A train to the tiny island neighborhood of Broad Channel and then either hike a half mile to the refuge or take a bus. When we visited the site we arrived by car via Cross Bay Boulevard after passing through the eclectic neighborhood of Howard Beach.

The refuge is a little bit of nature in the bustling cityscape. Although the refuge itself is relatively tranquil, it is not easy to forget where you are. Jumbo jets taking off and landing at JFK Airport screech overhead; the shoreline is strewn with styrofoam, plastic, old tires, and other manner of debris. On a clear day, the Manhattan skyline is visible to the north, but on our visit, the skies were hazy and all we could see in the distance were the silhouettes of three huge landfills that over the years have leeched noxious pollutants into Jamaica Bay. Because it is subjected to very little tidal flushing, the bay has been in poor health. It is apparently recovering from the most severe pollution that adversely affected water quality in the mid to late 1900s. Residents of the area are looking forward to the day when the bay becomes safe for swimming and the fish they catch can be eaten, but more work needs to be done before a clean bay becomes a reality. Despite the pollution, terrapins abound. They don't seem to mind the conditions and may be fairly tolerant of poor water quality in Jamaica Bay.

We met Amanda Widrig there on an unusually hot and humid day in June 2004. As soon as we entered the refuge and made our way down the main trail, we saw three female terrapins: one in the process of depositing eggs in a nest, a second digging a nest, and a third wandering across our path. We happened to arrive on a peak nesting day in New York, two to three weeks earlier than we expected to see any signs of nesters on Cape Cod. Widrig, who was preparing to enter a graduate program at Hofstra University, was working under the direction of Dr. Russell Burke. Lucky for her on this busy day, she had a small group of dedicated assistants that included a new Jamaica Bay Wildlife Refuge volunteer, a high school student, and a Central Park zookeeper who was helping out on his day off.

From the number of female terrapins loitering in the offshore shallows, the mounds of terrapin eggshells, and the number of depredated nests, it was apparent that a significant number of female terrapins elect to nest on Ruler's Bar, where Burke and his students find approximately 2,000 nests each year. This is a phenomenal number of nests compared to many other terrapin nesting sites, potentially making Jamaica Bay a very productive terrapin hatchery. However, there has been a recent introduction to the habitat that has upset the ecological balance. Since the 1980s, when Bob Cook was a natural resource management specialist with Gateway National Recreation Area and first chronicled the activity of terrapins in JBWR (Cook, 1989), this colony has faced a new challenge: raccoons. Sometime within the last twenty years, these predators have made their way to Ruler's Bar. There is considerable speculation regarding their mode of arrival. No one knows how raccoons found their way into the refuge. It is certainly possible that they traveled via Cross Bay Boulevard on the bridges connecting the island to Howard Beach or Rockaway. Raccoons can swim, so perhaps the predators arrived by water. One anecdote describes the efforts of a well-meaning animal control officer who released the raccoons after capturing them in local residential neighborhoods. Russ Burke estimates that the contents of over 90 percent of nests become meals for raccoons. The raccoons that we saw during our visit were very bold. They scampered around in broad daylight looking for newly laid terrapin nests. The boldest raccoon of all was actually digging out eggs as a female terrapin was laying them. Our presence caused the raccoon to scamper away, but when we peeked down to assess the damage, we saw the female terrapin's carapace covered in egg yolk (plate 7). Eggs with claw marks were scattered over the area—another failed nest at Ruler's Bar. Although we may never know how the raccoons arrived, they are certainly taking their toll on terrapin eggs. Some raccoons have even feasted on adult female terrapins on nesting forays.

Burke is studying nutrient flow in JBWR and the contribution of terrapins and their eggs to the food web. Of all the terrapin populations I have visited, JBWR provides the most unlikely habitat for the species. The salt marsh is not lush or extensive, the water is polluted, and a very dense human population resides in close proximity. Aside from the refuge itself, there is no other nesting habitat. Houses, apartments, storefronts, busy roadways, and artificially constructed basins surround the area. If the refuge had not been acquired by the National Park Service, it would be easy to predict that the terrapins would have disappeared years ago.

After the day's surge of nesting subsided, Widrig and her volunteers still had an evening of work ahead of them. There were a dozen new nests that were to be protected with mesh screens after the eggs were counted, weighed and reburied. Of twelve females that were captured after laying their eggs, only two were already marked with special electronic devices known as PIT (passive integrated transponder) tags. The remaining ten were to be processed and marked. The sun would set before the team would finish for the day.

Scattered across the north and south shores of Long Island, the other diamondback terrapin clusters have not been extensively studied, and very little is currently being done to assess the status of the populations.

## Cape May

The road to Cape May, New Jersey, takes the traveler to the Garden State Parkway, a ribbon of road that becomes scenic after one drives south through the Woodbury tolls. The Garden State is a pleasure to drive during the winter, but it becomes a bumper-to-bumper nightmare during weekends in the heart of the summer. Everyone is heading toward the New Jersey resort communities consisting of ocean-bordering towns, otherwise known by New Yorkers and New Jerseyites as simply "The Shore." For most of the trip south, the Garden State Parkway meanders near the New Jersey coastline, where occasionally one can see stretches of salt marsh east of the parkway. Every so often, the scenery is punctuated by steeply rising industrial complexes that spew dark clouds into the sky, oil refineries and power plants that are processing crude oil and producing energy.

The New Jersey salt marshes are extensive. Acres and acres of huge, flat expanses of salt marsh fill shallow land masses between the mainland and barrier beach islands that extend along the New Jersey coast from the Hudson River estuary to Delaware Bay. New Jersey barrier islands were used by Native Americans as warm weather outposts and fishing camps. When the weather turned colder, the natives would make their retreat from the shoreline and return to inland encampments. It wasn't until the 1600s that European settlers began to populate the islands. The Jersey shoreline morphed into a resort location in the 1800s, when it was hyped as an area that featured sea bathing and seafood. Many of the barrier beaches formerly consisted of sand dunes, but these have been bulldozed, extensively developed, and flattened to enhance ocean views. The coast has been buttressed with seawalls and fortified by small jetties, called groins, thus providing a prime example of human-engineered impacts on barrier island evolution.

Barrier beach areas are accessible by causeways where traffic zips back and forth during the busy summer months. The causeways span channels that are used by fishing vessels and recreational boaters. There is constant pressure to keep the channels open and to make them deep enough for navigation. This necessitates dredging and finding a home for the fill removed by dredging operations. These are the same channels that may be used by hibernating terrapins during the winter months. It is in this developed landscape that terrapins can be found. Terrapins have been known to dwell as far north as Sandy Hook and as far south as Cape May, the southern tip of New Jersey. They are found mostly in the channels, sounds, and other waterways within the large marshes between the mainland and the barrier islands, within sight of high-rising hotels and condominiums. In the 1970s, studies were conducted by Joanna Burger and her colleagues at Rutgers University on a diamondback terrapin cluster found on Little Beach Island, part of the former Brigantine National Wildlife Refuge in Barnegut Bay (Montevecchi and Burger, 1975). The current distribution of terrapins in this part of New Jersey has not been ascertained.

Most of the research, education, and conservation of New Jersey diamondback terrapins is being conducted at the Wetlands Institute in Stone Harbor, on the apron of a salt marsh between the mainland and the barrier beaches in Cape May. The Wetlands Institute is a private, nonprofit organization, founded in 1969 to promote appreciation of wetlands and coastal ecosystems. Roger Wood, who is also a professor at Richard Stockton College of New Jersey, is director of research at the Wetlands Institute and has been studying terrapins for many years. He led Earthwatch terrapin research expeditions in Florida in the early 1980s and has supervised diamondback terrapin projects at the Wetlands Institute since 1989. Under his direction, the Wetlands Institute runs summer programs, trains interns, and conducts educational sessions that focus mainly on Cape May terrapins. In a novel program sponsored by the New York Turtle and Tortoise Society, scholarships are awarded to Asian conservationists that allow them to assist in terrapin conservation efforts during the busy summer field season.

Wood and his assistant Christina Watters hosted my midwinter visit. With record low temperatures, I had no false hopes of seeing Cape May terrapins in the wild in January. I was, however, interested to see captive terrapins, many of which had been injured and were undergoing treatment and rehabilitation. Some were spending the winter at the animal facility at Richard Stockton College. Most of the adult terrapins were females. A few had fiberglass bandages that were holding together parts of their shells (fig. 1.7). Clearly, these terrapins weren't looking both ways when they crossed the road. Some were healing nicely and were destined for release in the spring. Others were so severely injured that they would probably be permanent residents of the Wetlands Institute's terrapin exhibit. The number of terrapins in the Cape May marshes is not known but is suspected to be relatively large, since approximately 500 females are found as road kill each year and that number does not seem to be declining. The ratio of females to males is not known with certainty since most of the observed terrapins are nesting females.

Major threats to terrapins in Cape May include automobiles and crab traps. Female diamondback terrapins are hit by cars when they make their way on shore to look for nesting areas. Because males and juvenile terrapins are attracted to crab traps, they enter the traps and are unable to escape. This results in drowning, a serious problem wherever terrapins coexist with blue crabs. Wood and his colleagues have been leaders in efforts to prevent road kills and crab trap drowning. Their efforts are described in chapter 6.

Just south of Cape May, the terrapin population in an area of Delaware Bay was studied in 1979. Terrapins were captured by trawling along 0.9 kilometers of Canary Creek, near the University of Delaware marine field station (Hurd, et al., 1979). Based on mark/recapture studies, 1,655 terrapins were in the creek in June of that year, but the number decreased to 378 in August, possibly reflecting estivation or the late summer dispersal that is seen on Cape Cod. In the Delaware marshes, which were predominantly populated by *Spartina alterniflora*, researchers counted 1.8 terrapins per linear meter of creek and found, from fecal analysis, that the blue mussel, *Mytilus edulis*, was an important constituent of the diet.

## Chesapeake Bay

Although only Maryland and Virginia geographically embrace Chesapeake Bay, the waters that drain into it also come from Delaware, New York, Pennsylvania, West Virginia and the District of Columbia. The watershed comprises 64,000 square miles threaded by over 150 major rivers and streams and hundreds of thousands of smaller ones. The bay, which assumed its current dimensions about 3,000 years ago, and the marshes along its shore are a rich habitat that was utilized by Native Americans for thousands of years. Students in the United States are often introduced to American colonial history by learning about John Smith and the first permanent European settlement at Jamestown in 1607. The early settlers had frequent, although not always peaceful, interactions with the natives, primarily Algonquians. Although many of the colonists did not survive past the first year after the settlement was established, there was a steady stream of new settlers and the replacements from Europe spread throughout the Chesapeake Bay region. Thus began modern human history in Chesapeake Bay.

The bay itself is a 320-kilometer (200 mile) long finger of water, poking into the city of Baltimore. Its narrowest region is a mere 5.5 kilometers (3.4 miles), but it widens to ten times that size at the mouth of the Potomac. Its 18 trillion gallons of water are held in a shallow basin with an average depth of 6.4 meters (21 feet). A few deep troughs allow navigation by large vessels. The bay's undulating shoreline has a length greater than that of California, Oregon, and Washington state combined. About half of the bay's water is fresh and comes from the watershed areas, mostly from the northern and western areas of the drainage. A small amount of the fresh water, approximately 10 percent, comes from the eastern shore. Half of the bay fills with water from the Atlantic Ocean. Thus salinity in the bay varies greatly with location and weather patterns; in general, the bay gets less salty as one travels southwest to northeast.

The Chesapeake shoreline provides an ideal habitat for terrapins, especially in locations where sandy beaches, which can be used for nesting, are fringed by salt marsh. Compared to other parts of their range, terrapins can be found in relative abundance in Chesapeake Bay, but there are no reliable records of their historic distribution. It is certain that their numbers dramatically declined from the end of the 1800s to the early 1900s, when a thriving commercial fishery supplied turtles for the food trade. With the relaxation of fishing pressure, the population was assumed to be recovering.

The rich estuarine resources of the bay provided a livelihood for the famous Chesapeake watermen, those who make their living by harvesting fish, oysters, and crabs. Despite the large decline in terrapin fisheries in the early twentieth century, a handful of watermen still catch terrapins. It is legal to do so, and the industry is regulated by the state of Maryland Department of Natural Resources. However, there are very few watermen who apply for licenses to fish for terrapins. All terrapins caught by commercial license holders must be reported, and recent annual harvest has been estimated to be 10,000 terrapins. Today, many terrapins are also caught as by-catch in fishing nets and crab pots and may go unreported. It is ironic that the terrapin fishery still exists in Maryland, but the enterprise is completely banned in the areas of the bay that constitute Virginia waters.

Even though the abundance of terrapins in the bay is well noted, the actual population status and distribution throughout the bay are not known. The disappearance of terrapins in some creeks and embayments in which they have been traditionally found has been attributed to fishing, unintended loss in crab pots, and loss of nesting habitat. The most thoroughly characterized cluster of Chesapeake terrapins inhabits the Patuxent River on the western side of the bay, where Willem Roosenburg of Ohio University has studied the cluter since 1987 and has captured over 8,000 individuals.

To address current conservation issues that may impact Chesapeake terrapins, it is very important to collect some baseline information about their current status. To this end, in 2001, a diamondback terrapin task force whose work is supported by an appropriation from the Maryland Congress, was created by the governor of Maryland. As part of the task force, a U. S. Geological Survey (USGS) team is currently heading a large-scale study of historic as well as current distribution of terrapins in the bay, population genetic structure relationships, nesting behavior and reproduction of terrapins throughout the bay. It will be important for the task force to delineate the habitat requirements for the species and to get a sense of the mortality associated with crab pots so that conservation recommendations can be formulated.

## The Carolinas

The most recent dedicated effort to study diamondback terrapins in North Carolina has been conducted by Kristen Hart. Hart completed a master's thesis on terrapins and continued to study the ecology of the species in North Carolina for her doctorate. She has documented four separate terrapin populations between Cove and Pimlico sounds (Hart, 2000). Hart has also worked with local crabbers in attempts to understand how bycatch of terrapins in crab pots will affect the status of the North Carolina populations.

The marshes in South Carolina share many features with northern counterparts. However, about 15 percent of South Carolina coastal marsh areas have been transformed into impoundments, surrounded by earthen dikes. These structures were originally constructed to cultivate rice, then later utilized to attract waterfowl.

Jeff Lovich, of the USGS, Whit Gibbons, from the University of Georgia, Savannah River Ecology Laboratory, and Anton (Tony) Tucker, currently program director of Sea Turtle Conservation and Research at Mote Marine Laboratory, are leaders in the field of herpetology. They have been pioneers in conducting research on turtles within an ecological framework, paying keen attention not only to the turtles but also to the habitat in which they live. They have been tracking the status of diamondback terrapins in South Carolina since 1983. Over a twenty-year time period, they have documented the decline in the population of *Malaclemys terrapin centrata* in the marshes around Kiawah Island, South Carolina. The terrapins at Kiawah Island generally have been found in the same creeks and same small reaches of water, generally within 100 meters (a bit over 100 yards) of previous captures. Some females have been observed to make 5-kilometer (approximately 3 miles) trips to nesting areas. Of 442 recaptured turtles, only twenty-five had moved from one area to another nearby creek. The terrapins stuck close to home after monstrous hurricanes, such as Hugo, which battered the area with 140 mph winds in 1989 (Gibbons et al., 2001). These destructive hurricanes hit the South Carolina coast every five to six years. Terrapin site fidelity, which appears to be true of all terrapin clusters, poses serious problems if the same creeks are used for trapping crabs. Terrapins won't move to another location where there would be less danger of being trapped and drowned in a crab pot.

The most recent survey of terrapins at Kiawah Island, in 2003, pointed to the severe decline in the number of turtles that could be captured. Using the same methodology and same degree of effort in one area where 200 terrapins were captured years before, only fifty could be found. There also has been a dramatic decline in the capture of younger turtles; some age classes could not be found at all. The reasons for the decline are complex and not completely understood, but several possibilities have most likely contributed to the diminishing numbers within creeks as well as several extirpations from specific creeks: overharvesting, loss of habitat, decline of health of the marsh, increase in the mink population due to reintroduction, increased beach use, and crab traps (Gibbons et al., 2004).

The hazards of crab traps are very serious in the Carolinas. In a study conducted from April to May of 1982 by Bishop (1983), creeks with known terrapin populations that were used for commercial crabbing were monitored. It was estimated that 2,500 terrapins were caught per day by 743 commercial crabbers in Charleston. As in the Kiawah Island marshes, the terrapin population in Charleston Harbor has declined since the 1980s. Within sight of Fort Sumter, the iconic site of the start of the Civil War, the current status of diamondback terrapins is being assessed by a team at the College of Charleston who work with South Carolina's Department of Natural Resources. Using sonic transmitters to track adults, David Owens and his students hope to find out how the terrapins use the harbor and why their numbers are declining. The team is mapping spacial and temporal habitat use on a fine scale and have identified mating aggregation, foraging, and nesting areas (Estep, 2004).

## Florida

Florida is home to more subspecies of diamondback terrapins than any other state. Five of the seven subspecies can be found in the Sunshine State. *Malaclemys terrapin centrata* extends from its range in South Carolina to northeast Florida near Jacksonville. *Malaclemys t. tequesta*, named after a Native American tribe, has been witness to over forty years of space mission lift-offs in its location off Merritt Island near the John F. Kennedy Space Station. This population, found in Merritt Island National Wildlife Refuge and Canaveral National Seashore, was studied from 1977 to 1979 and again from 1992 to 1993 (Siegel, 1993). In this protected habitat containing three large brackish lagoons on Florida's Atlantic coast, a considerable long-term decline in the population of terrapins was observed. In the late 1970s, Seigel estimated the population to number in the hundreds: 404 at one site and 213 at another. The gender ratio was skewed and ranged from 5:1 to 10:1 females to males, depending on the month of the year that sampling was conducted. Aggregations of terrapins were common from March to mid May. During that period, it was noted that raccoons were predators of mature females as they came on land to lay their eggs. At the same site, using the same sampling method, only one terrapin was observed in 1986 and none in 1993. This represents an extirpation without an obvious cause. In a 1998 ranking of vulnerability for extinction of various species at the John F. Kennedy Space Center, the Florida East Coast terrapin was listed as one of six species that were categorized in the highest priority in need of conservation. The other endangered species were the Eastern indigo snake, Southeastern beach mouse, Florida scrub jay, Atlantic green turtle, and the manatee (sea cow) (Breininger et al., 1998).

Joseph Butler has studied the status and distribution of the Carolina diamondback terrapin (subspecies *centrata*) in Duval and Nassau counties in the northeast corner of Florida, where the terrapins inhabit over 50 nautical miles (about 95 kilometers) of tidal creeks and marsh adjacent to the Intracostal Waterway. In these creeks, *Spartina alterniflora* predominates, and animals include bottlenose dolphin (*Tursiops truncatus*), manatee (*Tichechus manatus*), and river otter (*Lutra canadensis*). Raccoons are abundant. Terrapins become active in this locale and are only seen when the water temperature climbs above 19°C (66.2°F)(Butler, 2000). Butler and his colleagues experimented with different terrapin capture methods for their surveys. They raked tidal debris to look for hatchlings and juveniles with little success. In the water, they used turtle hoop traps, gill nets, otter trawls, and modified crab traps. Although they were able to capture elusive terrapins with all methods, some techniques worked better than others. Their study highlights the fact that methods that work well in some areas may have limited usefulness in others. For example, otter trawling is efficient as a method of terrapin sampling in some areas, but where oysters or oyster shells are found in abundance on the bottom, nets can be torn to shreds. Cast netting only works well if terrapins are congregated in a small area and if someone has skill with the technique.

During spring break in 2003, I traveled with my husband Nick to the Florida Keys and Everglades National Park. I was in search of the mangrove terrapin. These turtles are so elusive that the casual observer would never have the opportunity to come across a specimen. We were scheduled to meet Brian Mealey and Greta Parks at mile marker 99, where we would find the Key Largo Ranger Station, part of the National Park Service, housed in the remains of an old, rather run-down motel on U.S. Route 1. Mealey has been following the terrapin population in Florida Bay for over ten years. He was finishing up a doctoral program at Florida Atlantic University and recently started his own environmental consulting firm. Parks, who works as coordinator of outreach programs at the Miami Museum of Science, assists Mealey with several projects. I had learned about their terrapin monitoring program from reading the abstracts from the Second Workshop on Diamondback Terrapins that took place at the Wetlands Institute in Cape May, New Jersey, in

2000. After I contacted Mealey, he generously agreed to let me tag along on one of his field trips into Florida Bay. He warned me that we might not see terrapins; it was a bit early in the season. But on a more positive note, he told me that terrapin spotting at this time of year was rather unpredictable and terrapins have been observed year-round. Still, he could offer me no guarantees.

Nick and I arrived shortly before our appointed 10:30 A.M. meeting, a very civilized time of day for field biologists. As I observed the clouds thickening overhead and the choppy seas caused by 20-knot winds from the northeast, I felt some trepidation about the prospect of heading out to the mangrove islands in Florida Bay. I have a long history of experiencing motion sickness, sometimes in relatively calm water just from floating up and down in the swells. I have always preferred to have two feet planted firmly on the ground. The sight of Mealey's boat being trailered into the ranger station did nothing to allay my fears. It was a small fiberglass-bottom craft with inflatable sides and a very shallow draft. We were certain to be knocked around a bit once we hit the open bay. After introductions were quickly made, I lost all thought of physical consequences and looked eagerly to the adventure that lay ahead.

The first matter of business was to check a number of osprey nests that Mealey and Parks had been monitoring. We would look for terrapins as we made a loop back to the dock later in the day. The osprey nests we were to visit were known to have either eggs or chicks, based on observations that had been made a few weeks prior to our trip. The boat proved to be the perfect vessel for navigating among the mangrove islands in the shallow reaches of Florida Bay and closing in on the osprey nests in stumpy mangrove trees. The large osprey nests resembled snarled masses of matted hair. While Mealey maneuvered the boat, Park used an extendible pole, topped with an auto rear-view mirror, to peer into the nests and check on the chicks. Several nests had "failed": They were abandoned and showed no signs of chicks. In others, turkey vultures were working on chick remains. After a bleak start, we came across several nests with one to three healthy chicks and doting osprey parents nearby. In cases in which chicks were sufficiently mature, Mealey and Parks quickly and expertly retrieved them from nests, one at a time, decorated their legs with identification bands, and took blood samples for chemical and genetic analysis.

After all osprey nests had been checked, we made a loop back to the north side of Nest Key. Mealey knew of a sheltered spot where terrapins had been known to congregate. The sound of our motor brought the curious turtles to the surface, but they were able to quickly dive out of sight and out of reach.

We anchored in the northeast corner of Nest Key and made our way upland of the red mangroves, *Rhizophora mangle*, that rimmed the little island and projected upward from their masses of knobby-kneed prop roots. It is the red mangrove that gives the subspecies of mangrove terrapin its name, *rhizophorarum*. Parks, wearing diving boots, quickly began to search around the base of a huge black mangrove, *Avicennis germinans*, anchored in an extensive mud plain. The black mangrove is easy to distinguish from the red mangrove; in place of prop roots, the black mangrove sends vertical roots, called pneumatophores, up from the soil (plate 8). These structures provide for gas exchange with the underground root system and look like a field of dried sticks, up to a foot long, inserted around the base of the tree. The pneumatophores are packed so tightly that in order for terrapins to navigate at the base of the trees, they must tilt their bodies to a vertical position and inch along sideways like a crab to make any headway (Wood, 1992).

In short order, Parks found a large female terrapin. When I attempted to follow in my aqua shoes, I immediately appreciated the value of the diving boots. As I sank thigh deep into the pale, mucky ooze, my water shoes were sucked off my feet. Nick wisely remained close to shore to tend to the boat. As a barefoot explorer, I took some photos and tried not to stay too long in one spot for fear that I would disappear into the soft substrate. Parks and Mealey found two other females in close proximity, partially buried in the muck, and I was assigned the task of washing them off so that they could be processed. In all the excitement, one female made her escape. Of the two that were left, one was a recapture but the other female had not been seen previously. Each weighed approximately 1,100 grams (2.4 pounds) and was as smooth as a river stone. They had "striped pants," vertical black markings on their hind legs. I have seen the striped pants on terrapins from New York and other sites. The pants become pantaloons after heavy rains as the terrapins strive to store water in the skin of their limbs to carry them over any spell of dry weather. The new capture was injected with an electronic tag, a PIT (passive integrated transponder) tag, and a blood sample was quickly and expertly taken by Mealey for genetic analysis. I noticed that these terrapins, found buried in the mud plain, were lethargic and easy to handle, most likely because they were experiencing some of the physiological effects of estivation. If they were Wellfleet terrapins, captured during the summer months, they would be scratching and chomping during weighing and blood collection.

Of all the habitats utilized by diamondback terrapins, the subtropical mangrove swamp is the most otherworldly. One would never equate this landscape to that of a salt marsh at the mouth of a river estuary. But Florida Bay is an estuary at the mouth of one of the largest rivers of all: a River of Grass, better known as the Everglades and beautifully described by Marjorie Stoneman Douglas (1947) and more recently by Ted Levin (2003). The slowly moving sheet of water that encompasses the Everglades is the river that feeds fresh water into Florida Bay, semi-encircled by the Florida mainland and the Florida Keys. Human activities have greatly changed the nature of the Everglades. Draining of land for agriculture and development by rechanneling and control of water flow has led to the shrinkage of the Everglades and a restriction of the flow of fresh water into Florida Bay. This has led to an increase in salinity in many parts of the bay. In most years, Florida Bay contains water that has more salt than the ocean. This is a condition referred to as hypersalinity. It is not unusual for some parts of Florida Bay to reach salinities that are twice the level of seawater! It is in this salty soup that tiny mangrove islands, called keys, poke their heads above the shallow waters. Some of these keys may be completely submerged at high tide. It is on and around such keys that the mangrove terrapins make their home.

The red mangroves that circle each key stretch their long prop roots, like tentacles into the briny water. A thin ribbon of dry land complements the red mangrove border. Parting one's way through the red mangrove thicket, the vista becomes a large, partially shrubby, partially muddy expanse. Black mangroves punctuate the salt pan landscape. It is among the vertically thrusting roots of the black mangrove where terrapins burrow in the mud year-round. Here they rest, protected from the heat and waiting for a dose of fresh water when they are not foraging in the nearby shallows. Atop these muddy expanses, rainwater will create shallow pools and terrapins can rehydrate. As the turtles store water, their skin expands. Periodically, the water evaporates and pools give way to quicksand-like ooze.

The mangrove islands and low-lying Florida Keys can be pummeled by strong winds and flood tides during hurricanes. Miller (2001) questioned whether hurricanes might cause *M. t. rhizophorarum* to redistribute. She studied the possible dispersal of terrapins as a result of Hurricane Georges, which hit south Florida on September 26, 1998, with sustained winds of 105 miles per hour and flooding due to tides that were 1.2 to 1.8 meters (4 to 6 feet) above normal. On small islands where recapture rates were over 70 percent, the female to male ratios varied from 21:1 near Key West to 5:1 in the Middle Keys. Miller used molecular genetic techniques to study the possible forced dispersal of terrapins as a result of the hurricane, reasoning that dispersal might introduce new genes into the subpopulations at different sites in Florida Bay and the Keys. This limited study did not find any evidence for changes in gene frequencies as a result of Hurricane Georges. Once again, terrapins have been found to stick close to home.

Active throughout the year, members of this southernmost subspecies feed on periwinkles that are found on the mangrove roots. They may also eat the smaller mangrove or coon oysters that adhere to roots. The inaccessibility of the mangrove islands to terrestrial predators such as raccoons is partially offset by the activities of the black rat, *Rattus rattus*, which has inserted itself into the mangrove community and may be a significant predator of terrapin eggs. Very few hatchlings or juvenile terrapins are ever seen.

We did not get to see the other reptiles such as water snakes and crocodiles that share the mangrove swamps with terrapins; Mealey was focused on the Florida Bay terrapins, which he has been studying since 1992. Over 800 of them have been captured and marked with PIT tags. By scanning the turtles with a wand, he can identify each specimen and determine when it was last captured. Most of the captured animals have been females, and Mealey has recaptured many of the same individuals during the years of his study. He estimates a 10:1 ratio of females to males in this Florida Bay cluster. Mealey has discovered that mangrove terrapins, true to the nature of their species, are homebodies. Most tagged females are recovered at the location where they were originally found. A few wander, such as the two females who traveled approximately three miles to another key.

It is not clear how the Florida Bay terrapins have fared over the years. The population is relatively inaccessible and has not been studied over a long period. The lack of coastal development on the mangrove islands and the protection of many of the islands as part of Everglades National Park have prevented some human impacts. However, the major anthropogenic changes that have occurred in the Everglades over the past 200 years has undoubtedly affected the mangrove terrapins, if only indirectly. It remains to be seen how the proposed restoration efforts in the Everglades will change the landscape for mangrove terrapins and their ability to utilize Florida Bay. This large-scale project should potentially freshen the water in the Everglades estuary. Presumably terrapins will experience less osmotic stress.

There are other locations where diamondback terrapins are found in Florida. Aside from Florida Bay and the Keys, *M. t. macrospilota* makes its home on the southwest and west coasts of Florida, while *M. t. pileata* can be found in salt marshes along the panhandle.

## Alabama, Mississippi and Louisiana

The Gulf Coast terrapins have not been continuously studied. Cagle collected specimens in the early 1950s along the Louisiana coast. He found *M. t. pileata* and *M. t. littoralis* and some terrapins that had morphological characteristics of both subspecies, suggesting intergradation. Cagle also reported an unusual gender ratio for the ninety-six terrapins he collected at Dulac, Louisiana: 4.4 males to 1 female (Cagle, 1952).

A more recent initiative to characterize the distribution and status of *M. t. pileata* was focused on Mobile Bay, Alabama (Nelson et al., 2000). The study was not successful in attempts to use crab-trap captures as an index of population size and structure because no terrapins were captured in traps during the period of the study. Some terrapins were found and collections of depredated nests were discovered in certain areas of the bay. The impact of road mortality on the population is being assessed.

## Texas

Prior to the mid-1990s, very little was known about the status of the diamondback terrapin in Texas. Historical reports suggested a sizable population, and the former existence of a commercial fishery was a matter of public record. *Malaclemys terrapin littoralis* has been found in various pockets along the Gulf coast of Texas, and some idea of its distribution was known from reports garnered from fishing boat operators, game wardens, and coastal fisheries biologists. In 1997, terrapin scouting trips were made to Lavaca, San Antonio, and Nueces Bays. Only Nueces Bay, near the Mexican border, had a sizable population.

In the late 1800s, a commercial terrapin fishery was active in a bayou near Galveston. Today, the bayou is part of a preserve comprising 2,800 acres of watershed that has been subjected to a number of stressors, including urban development and cattle grazing. Close to 95 percent of the marsh vegetation has disappeared, and increasing water levels threaten to erase the marsh completely. The most recent attempt at comprehensive assessment of the status of diamondback terrapins near Galveston was conducted during 2001 to 2002 by the USGS team led by Jennifer Hogan. The team worked with the U.S. Fish and Wildlife Service. They captured 135 terrapins in crab traps offshore and in lagoons, and by hand as terrapins walked on shore. Oyster reefs constituted a considerable dimension of their habitat. Hogan was able to document the first terrapin nest reported in Texas since the 1960s; it was the only one found during two seasons of study. The survey was conducted in a limited area and focused on South Deer Island, a small island in Galveston Bay (Hogan, 2003). There is quite a bit of Texas coastline that has yet to be thoroughly surveyed for the presence of diamondback terrapins.

## The Ideal Habitat

What determines the ideal habitat to sustain diamondback terrapin populations? The answer to this question is not simple. Throughout their range, the similar habitat requirements include calm, brackish waters near salt marshes or, in some cases, mangrove swamps. But they are not found in all calm brackish waters near salt marshes or swamps. They can tolerate variable salinity; they are found with variable proximity to salt marshes, and the salt marshes are of variable size. So what do they specifically need in order to be successful? One report (Palmer and Cordes, 1988) used only one parameter, the availability of suitable upland nesting areas, to generate a model for habitat suitability. Nesting areas are critical for sustaining diamondback terrapin populations. So perhaps the variety and type of upland nesting areas must also be factored into the equation.

All stages of terrapin life history must be considered in defining habitat requirements. For example, hibernation sites may be different from those areas used for foraging, mating and other activities. Without suitable hibernation space and substrate, terrapins will not survive winters in parts of their range. And what about the younger terrapins? Those that are newly hatched, juvenile, and subadult may have some habitat that overlaps with adults, but early terrapin years may be predominantly spent in drier areas of the marsh and in marsh uplands (chap. 4). It is therefore also important to consider the quality and quantity of the coastal zone surrounding the marsh and to identify those aspects of habitat that will maintain and protect younger terrapins during their most vulnerable period.

It may also be true that anthropogenic factors unrelated to natural habitat, such as commercial and recreational activities, have a profound impact on the distribution of the species in any particular location, an aspect that is further explored in chapters 5 and 6.

# Chapter 3

## Reproduction: Insurance for Species Survival

R EPRODUCTION IS ONE of the fundamental processes that characterize all living things. Every individual organism, whether it is a single-celled bacterium or a large vertebrate, has a finite life span. Reproduction serves to pass on the traits of individual members of the species to the next generation and thus ensures the survival of a species. Individual turtles have long life spans that factor into their reproductive strategy.

Turtles, like birds, are oviparous. This means that fertilization of eggs occurs internally but the eggs are deposited externally to complete their development. Turtles do not experience metamorphosis or pass through complex life cycles. Turtle hatchlings are miniaturized versions of the adults. Aside from growth, the only complex morphological changes they experience occur during sexual maturation.

Diamondback terrapins have developed a reproductive strategy that shows some subtle variations throughout the range of the species. Some of our knowledge about terrapin reproduction comes from early observations during the time period that terrapins were under cultivation, but there are increasing numbers of reports describing terrapin reproduction in wild populations. Predation of offspring will influence reproductive success. Therefore, most species face a trade-off: large numbers of offspring that are small and not well developed, or smaller numbers of larger, well-developed offspring, which may have a better chance of survival. In all types of turtles, the high rate of predation on eggs is an important factor that shapes the life history of the species. If high percentages of eggs are destroyed by predators, there is a better chance of passing on the genetic potential of the species if the clutch size is large.

However, a female turtle can only harbor a certain number of eggs. Each large female sea turtle may be able to produce more than a hundred eggs per clutch, but the smaller size of the diamondback terrapin will limit her clutch size to considerably fewer eggs. Although diamondback terrapins fall in the middle of the turtle reproductive spectrum in terms of clutch size, they are similar to other turtles in having well-developed but small and vulnerable offspring. Another factor that makes a positive contribution to reproductive success for some species is the investment of parental care. Turtles are deadbeat parents—their offspring are completely on their own once they emerge from their eggs.

The strategy for the survival of a species is next shaped by another trade-off: mature early, within a short time span while the animal is still small and at risk from predators, or spend the early years growing, mature later, and reproduce at a later age. The former strategy is geared to ensure a shot at reproduction before death by depredation; the latter strategy focuses on attaining a size that makes the animal immune from predation but delays reproduction. Turtles have chosen to delay reproduction until they are relatively predator proof. In the early years, their energy is devoted to growing their coat of armor so that when they reach maturity, they will live long and reproduce often.

## Sex Ratio

Successful reproduction in turtles mandates that there are males and females in a population. However, not all diamondback terrapin populations have equal representation by the sexes. It is intriguing to speculate about the nature of the biased adult sex ratios for various clusters of diamondback terrapins. In some cases the number of mature males is approximately equal to the number of mature females, but in most locations sex ratios are biased toward females. A few sites appear to be exceptions, with males predominating. Before we try to understand the significance of the sex ratio for the sustainability of diamondback terrapin populations, it is reasonable to question the possible causes and the validity of published sex ratios.

The sex ratio in an adult turtle population may be partly explained by the sex ratio of hatchlings. The terrapin exhibits temperature-dependent sex determination (TSD). The sex of the terrapin hinges on the incubation temperature of the egg from which it emerges. Biased ratios of adults within a population may reflect disproportional sex ratios of hatchlings due to the nature and hence temperature of nesting sites in the various clusters. It is impossible to distinguish hatchling males from hatchling females with complete confidence based on external morphology. To determine the sex of hatchlings, investigators must use invasive techniques, such as laparoscopy, or sacrifice the turtles and perform histological examinations by looking at gonadal (sex organ) tissues under a microscope. Thus it is difficult to correlate incubation temperatures in natural nests with sex ratios in the emerging hatchlings.

Disproportionate numbers of males or females may also be attributed to differential mortality. There is no obvious reason why there should be differential mortality in wild hatchling or juvenile terrapins. For adult and subadult terrapins, other factors may affect life span. Depending on design of crab traps, males and females may suffer differential mortality rates as a result of accidental entrapment and subsequent drowning. Crab pots tend to trap more males than females (Roosenburg and Kelley, 1996) and can thus skew populations ratios toward a female bias. Because of their time spent on land, females are predictably more at risk of predation than males. In some regions, adult females are particularly vulnerable as victims of automobiles when they cross roadways to nest. With the expansion of the costal road network to service resort and vacation areas, females in a population will be disproportionately killed as a result of road motality. A study by Gibbs and Steen (2005) showed that sex ratios in turtle populations become more male biased where there are more roads and where females exhibit more terrestrial movement than males. For diamondback terrapins, the ratios in locations with high road mortality should show more sex bias toward males, which is not always the case. Since terrapins rarely roam, movement into clusters (immigration) or out of clusters (emigration) is unlikely to account for sex bias.

In some instances, reports of sex bias may be attributed to some degree to sampling bias. Female terrapins are easier to find because they spend more time on land. Small juvenile females can often be mistaken for mature males. Younger terrapins are almost impossible to find. And certainly, different seasonal sampling is expected to result in different sex sampling. This was reported by Seigel (1980a) in the Indian River, near the central Florida coast, when he showed that the winter sampling ratio of males to females was 10:1 while the late spring/summer ratio in the same location was 5:1. After mating, males may disperse and be more difficult to capture. Sampling in creeks adjacent to popular nesting areas may yield more females than males. Thus there are inherent problems when researchers attempt to ascertain the sex ratio of their terrapin colonies.

Lovich and Gibbons (1990) found a male biased population (average ratio of males to females was 1.78:1) at Kiawah Island, South Carolina. They used a variety of collection techniques and equipment, including trammel nets and seine nets. Under conditions that produced almost the same probability of recapturing individuals from either sex (0.44 for males and 0.38 for females), males always outnumbered females despite some yearly variations in the ratio. Lovich and Gibbons attributed this male bias to the fact that terrapin males mature at a much faster rate than females. Males mature in about three years and females take twice as long to reach adulthood, so one would expect more males in the adult population. Although this seems like a very reasonable explanation for a male-biased sex ratio, researchers at other sites are not finding the same male bias. A recent population sampling at Kiawah Island may point to one potential cause of biased sex ratios. Long-term studies at the same Kiawah Island creeks indicate a change in the sex ratio. It seems that crab pots are responsible for increasing the mortality of males, and thus the population sex ratio has shifted to the point where there is now a slight female bias (Gibbons et al., 2004).

What would be the ideal sex ratio to sustain terrapin populations? In studies in the early 1900s of farm-raised terrapins, the highest fertility was observed when the ratio of females to males was 5:1 (Hildebrand, 1932). No one knows the answer to the question of optimal sex ratio for terrapins in the wild, but most researches have the opinion that the loss of significant number of adult females can be the death knell for a population.

## Sexual Size Dimorphism

As outlined in chapter 1, there is a dramatic size difference and age in reaching maturity between adult male and adult female terrapins (plate 1). Adult females are always much larger than males and require more time to mature. Secondary sexual characteristics also distinguish the sexes: Females have larger heads than males; males have thicker tails with the cloaca positioned outside the posterior margin of the plastron. Why should diamondback terrapin females be so much larger than males? The same phenomenon is true in some other turtle species, but in a few types of turtles, such as the snapping turtle, the male is larger than the female. If males display aggression and have to fight for mates, larger size would be an advantage. If females "select" mates, larger male size may be a factor in the selection process. Because male terrapins are so small, these factors are probably not significant in driving the size of males. It could be argued that it is an advantage to the species when adult males and females achieve vastly different sizes because the sexes are not forced to compete for the same food resources

Other size advantages are more obvious. If the male matures early, even though he achieves a smaller size, he increases the number of times he can mate throughout his lifetime, thus increasing his lifetime reproductive potential. If we assume that female body size has some correlation with clutch size; that is, the number of eggs that can develop and be laid at a given time, later maturation and attainment of a larger size would be an advantage by increasing the number of eggs per clutch and thus the female's lifetime reproductive output. The relationship between size of females and clutch size is explored later in this chapter.

## Courtship and Mating

Although some aspects of terrapin mating have been elucidated from captive breeding programs, one of the missing areas in terrapin biology is observation of mating in the wild. In the North, we know that the more obvious sightings of mating horseshoe crabs signal the advent of terrapin mating season.

Mating has been observed in a few natural terrapin colonies. Chesapeake watermen told of great concentrations of terrapins in specific creeks in the early spring that may have represented mating aggregations. Siegel (1980c) reported mating aggregations at his study site at Merritt Island, Florida. On Cape Cod, an annual mating aggregation has been observed in one small cove in Wellfleet Harbor during late spring, although mating terrapins are sighted occasionally throughout their activity period. Aggregations make sense for mating in an aquatic species. It saves a lot of time for a seasonally active species in which males and females must search for a mate in a short time span. In Wellfleet, I have seen females being pursued by one or more males and it is often possible to scoop up a mating pair in a net from a kayak. In one scoop I once netted a three-year-old male and a precocious six-year-old female. Apparently, both turtles had recently reached maturity and were most likely mating for the first time or at least going through the motions. These ages are relatively young for terrapins to be mating on Cape Cod.

Even though terrapins have a distinct home range (the area they utilize for day-to-day and seasonal activities), males do not display territorial behavior. I have never observed males fighting for females. The whole scene is rather congenial, almost businesslike. It is still a mystery how and why the Wellfleet terrapins flock to this particular cove to mate. Other mysteries about the mating process abound. How is this annual event initiated? Who attracts whom, and are chemical/olfactory attractants or pheromones involved? How common is mating at other times of the year and at other locations?

In one of the few studies of diamondback terrapin mating in the wild, attempts were made to observe mating behavior from behind natural blinds in water that offered only 1.0 meter visibility. Twelve matings were witnessed in canals near Merritt Island during spring aggregations where six to seventy-five individuals were observed. All mating occurred in daylight when air and water temperatures were similar. Water temperatures ranged from 24.8 to 27.0°C (76.6 to 80.6°F) and air temperatures ranged from 22.8 to 27.0°C (73 to 80.6°F). The turbidity of the water prevented the entire courtship and mating ritual from being witnessed. From snapshots of the ritual from different mating pairs, observers could put together a scenario in which a female first floats on the surface, with the male approaching from the rear and nudging the female's cloacal region with his snout. Within a minute, the male mounts and copulation occurs immediately. Injection of sperm into the female is completed over the span of 1 to 2 minutes. If the female swims away before copulation, the male may follow in active pursuit (Seigel, 1980c). During May and June, in the clear waters of Chipman's Cove on Cape Cod, terrapin mating plays out in a manner similar to that described by Seigel (1980c). The entire courtship and mating sequence appears to be fast and furtive (Brennessel and Lewis, personal observations). Only continued observations of mating in the wild will help to identify any courtship rituals or subtle behaviors that are important in the manner in which this event is orchestrated.

Female terrapins can store sperm for several years. Observation of females at terrapin farms revealed that some could produce eggs up to four years without contact with males. There was some indication that the ratio of fertile to infertile eggs may have decreased as a result of sperm storage (Coker, 1920). The obvious advantage for terrapins, as well as other species in which the phenomenon of sperm storage occurs, is that it is not imperative for a female to find a second or third mate if she will produce more than one clutch per year. Furthermore, a lack of males or low ratio of males to females in the population will not prevent the female from reproducing every year. The ability to store sperm would be an advantage in populations with female-biased ratios.

Because females can mate with more than one male each season and store sperm, there is the potential for each clutch of eggs to be fathered by more than one male. Several researchers are interested in whether a single clutch of eggs is the result of multiple paternity. With the possibility of multiple matings and the ability to store sperm, it is not clear how many males' or which male's sperm fertilize the eggs of each clutch. Molecular techniques are being used to address this question, and we will most likely find that some of the hatchlings in some clutches have different fathers. Using genetic tools, preliminary work in this area suggests that multiple paternity does occur but that it may not be as common in diamondback terrapins compared to other turtle species (Argyriou et al., 2004; Hauswaldt, 2004; Page and Brennessel, 2005). The possibility of sperm competition is related to the issue of multiple paternity. In many species with multiple matings, it has been shown that the sperm of one particular male may preferentially fertilize all or most of the eggs. When a female has mated with several males, the male whose sperm is utilized will have the best chance of passing on his genes to future generations. It may be a case of the survival of the fittest sperm!

## Nesting

The nesting activities of female terrapins are inarguably the best understood aspect of the reproductive ecology of the species. There have been many studies and observations of terrapin nesting throughout their geographic range (table 3.1). However, these studies may not be as easy to conduct as one would think. A researcher must be in the right place at the right time to observe terrapin nesting. Not only are terrapins elusive in the water; they conduct their nesting activities in a most secretive manner. When a female is on a nesting run, we must remain quiet and hidden if we hope to observe her through the entire spectrum of her nesting activities. If we are lucky and are observing in a sandy area, we may find tracks that lead us from the creek or marsh to a nesting terrapin or a completed nest (fig. 3.1). In vegetated nesting areas, even Sherlock Holmes would find it difficult to remain on the trail of a nesting terrapin.

Female terrapins are known to lay eggs more than once each season. In captivity, where mating, nutrition, and growth were optimized, up to five clutches of eggs per female were recorded in one season (Hildebrand, 1932) although two to three clutches a year is more common. Clutches are separated by approximately 14 to 17 days, the length of time it takes for development of a set of eggs. Once she is ready to lay her eggs, the female must leave the relative safety of the water and trudge onto the land to dig a nest and deposit them. The trip has many perils. In some cases, the female may face vertebrate predators. Raccoons in Jamaica Bay Wildlife Refuge have been known to kill

*Fig. 3.1. Tracks, disturbed sand, and the tracings of a tail indicate a terrapin nesting area.*

and eat adult female terrapins when the females travel over land to nest (Feinberg and Burke, 2003). Long journeys on land also bring the danger of dehydration and overheating. Some females make the trip to nesting areas several times before conditions are right and they deposit their eggs. These forays are accompanied by sand sniffing, seemingly random digging with the snout and forelimbs, and even digging of a nesting chamber. But for reasons that are unclear, the female may abandon the nest before she deposits any eggs (fig. 3.2). It is fairly common, especially in the North, to observe aborted nesting attempts in which gravid females trudge onto land, scout around for a proper nesting site, and then about-face and return to the marsh. In some cases, this type of aborted nesting may be attributed to human activities: Noise, traffic, bicycles, pets, and other types of commotion may chase nesting terrapins back into the marsh. Peter Auger's group (Auger and Giovannone, 1979) observed certain females attempting to lay eggs five to six times over a period of a week to ten days. Each attempt was accompanied by an anthropogenic disturbance that drove the female back into the water. This nesting delay may actually be detrimental to the population, because pushing nesting to later in the summer may delay hatchling into late fall, especially if a second clutch is involved. In the North, this poses problems for hatchlings when the temperature drops in October and November.

Human activity is not the only cause of aborted nesting. I have witnessed, time after time, a gravid female begin to prepare a nest and abandon it, only to move on and repeat her attempted nesting many times before she actually deposits eggs or calls it a day and heads back to the marsh . . . only to try again in a day or two. The area she digs up often looks like a battle zone, pocked with holes that are the remains of aborted nests. When I dig down into some of the abandoned nest cavities, I occasionally find a large rock or thick plant root that most likely signaled the female to try another spot. Sometimes there is no obvious reason why a nest has been abandoned. Perhaps the female is trying to confuse predators. Females also make nonnesting excursions. I have sometimes found nongravid females on land, rooting about as though they were preparing to deposit eggs but never getting down to business.

In some locations, human activities may actually shift the pattern of nesting from diurnal to nocturnal. Although nesting in daylight is the norm in most terrapin colonies, on Sandy Neck, Barnstable, Massachusetts, (Auger and Giovannone, 1979), Little Beach Island, New Jersey (Montevecchi and Burger, 1975), Patuxent River, Maryland (Roosenburg, 1994), Jamaica Bay,

*Fig. 3.2. Aborted nest. A female terrapin began to dig out a nest but then abandoned it. There were several aborted nests in the immediate vicinity, all dug by the same female.*

New York (Feinberg and Burke, 2003), and in other locations, a significant amount of nesting activity may be occurring at night (table 3.1).

The weather is also a factor that determines nesting events. Air temperatures must be warm enough to power the movements of the nesting female but not so hot that she will overheat during a nesting foray. Feinberg and Burke (2003) have found the air temperature during optimal nesting times in Jamaica Bay to be 25.4 ± 3.2°C, (77.7 ± 6°F), while nesting did not occur when air temperatures exceeded 35°C (95°F). Peak nesting was observed at 25 to 75 percent cloud cover. In central coastal Florida, Seigel (1980b) observed that most nesting occurred under clear skies with an air temperature maximum for nesting at 36°C (96.8°F). Most investigators find that nesting does not usually occur under completely cloudy skies or during periods of rain. Perhaps female terrapins cue into the position of the sun to help with nest site selection. This visual cue may be important to position the nest in a location where the amount or angle of the sun will have a positive impact on nest temperatures throughout incubation.

Temperature restrictions may define the beginning and duration of the nesting season. The commencement of nesting for each colony varies predictably as a function of latitude as well as seasonal temperature variation. In the South, nesting may start as early as the beginning of May, while on Cape Cod, nesting rarely occurs before the middle of June and is sometimes delayed into the first weeks of July (table 3.1). In Florida, where the nesting season is longer, nesting and hatching may overlap; turtles in nests laid in May can be hatching while other nests are still being laid (Butler, 2000). With an extended nesting season, it is possible for southern terrapins to produce three clutches of eggs per year, while two is the maximum number of clutches for females in northern populations. Although double clutching has been observed on Cape Cod, it may not be the norm, especially when nesting does not begin until July.

In a study of reproductive ecology at Jamaica Bay Wildlife Refuge, Feinburg and Burke (2003) observed three distinct peaks of nesting activity separated by approximately two-week intervals. The first peak was characterized by large numbers of nesting females concentrated within a few days. During the following two nesting periods, fewer nesting terrapins were observed. In Wellfleet, three peaks of nesting activity are also observed but they occur over a more contracted time period. For example, in one year there were ten days between the first and second nesting peak and five days between the second and third. While it may be possible to speculate that the nesting time intervals observed at Jamaica Bay might correlate with multiple clutches from the same females, the intervals in Wellfleet nesting activity are not as easily related to multiple clutches and could possibly be related to variation in weather conditions.

There is a great variability in the distance females must travel on land in order to find suitable nesting habitat (table 3.1). On mangrove islands in Florida Bay, the nesting area is a few meters from the water. In Jamaica Bay, New York, nesting occurs about 180 meters (200 yards) or less from the water (Cook, 1989). When I visited Jamaica Bay during nesting season, I observed that the nesting areas were very close to the water. Female terrapins could be seen bobbing up and down within a few meters of the nesting beach, waiting for the opportune moment to come ashore. On Cape Cod, females sometimes make astounding trips across large expanses of marsh and sand dunes. Round trips of up to 1600 meters (about a mile) are not unusual (Auger and Giovannone, 1979). In Barrington, Rhode Island, nesting areas are not visible from the marsh. It sometimes requires long treks through thickly wooded areas before a female finds suitable nesting substrate.

In areas in which upland nesting sites are a considerable distance from the water and in areas with a great deal of tidal height variation, tides may have a significant impact on nesting activity. In the Northeast, such as at Oyster Bay, Long Island (Bauer and Draud, 2004), Sandy Neck, Massachusetts (Auger and Giovannone, 1979), Jamaica Bay, New York (Feinberg and Burke, 2003), and Barnegat Bay, New Jersey (Burger and Montevecchi, 1975), nesting occurs at all tidal heights but is concentrated around the hours of high tide. In northeast Florida, tidal nesting preference was found to be several hours before to one hour after each high tide (Butler, 2000). High tides float females up into the marsh, where they are closer to potential nesting sites. Not only does this mean that they have less distance to travel to high, dry nesting areas, but it allows them to have a gauge for the high tide line and thus a location to dig a nest that will be safe from tidal inundation. In some areas, a significant increase in nesting activity is observed at the time of the full moon, when tides are at their highest (Bauer and Draud, 2004). Although high tide nesting is common, some females nest to the beat of a different drummer and are found on nesting runs at mid and low tides.

## Nest Site Selection

It is not easy to decipher all of the instinctual factors that drive females to select the exact time and place for nesting. In most terrapin colonies, females display a preference for sandy, nonvegetated areas. Sandy soil of loose particle size may be optimal for gas diffusion and may be necessary for proper embryonic development. Sometimes sites are used in which the sand is relatively compact or strewn with gravel or shells. Large sandy areas offer less shading, and thus result in higher soil temperatures. Lack of nearby vegetation lessens the probability of destruction of eggs by plant roots. The disadvantages to open sandy areas include the higher probability of desiccation of eggs and chance of erosion due to wind. In contrast, nests in vegetated areas are more prone to root infiltration and are subject to cooler incubation temperatures. Terrapins are limited in terms of nesting sites and substrate type, depending on their location. On mangrove islands in Florida Bay, the nesting area is limited to a narrow margin of sandy marl between the hypersaline water and the swamp. In Chesapeake Bay, the sandy stretches around the bay are narrow, discontinuous, and interspersed with sections of salt marsh. In Cape May, New Jersey, females utilize the causeways to reach limited sandy areas that border the roads.

Terrapins may nest in clusters, with many females using the same small stretch of sandy terrain. This is especially true where nesting habitat is limited by natural topography or in areas populated by humans where much of the historical nesting habitat has been converted to home sites, driveways, and roadways. In many cases, sandy areas have been "hardened" with asphalt and other materials to allow the passage of automobiles. Seawalls and revetments may prevent females from utilizing some potential areas. Auger and Giovannone (1979) reported nest density in Sandy Neck, Barnstable, Massachusetts, to be fifty nests per 96 hectares (237 acres), which is relatively disperse compared to reported densities of 289 nests per hectare (2.4 acres) in northeast Florida (Butler, 2000), 220 nests per 1.4 hectares (3.5 acres) near Barnegat Bay, New Jersey (Montevecchi and Burger, 1975), and 446 nests per hectare (2.4 acres) in a nesting area within a Rhode Island wildlife refuge (Goodwin, 1994). Even within the same habitat, such as the Patuxent River, nest densities may vary from 60 nests per 0.25 hectares (0.6 acres) at one location to 225 nests per hectare (2.4 acres) at another location (Roosenburg, 1994). It is not clear whether nest densities in some localities are due to local colony size or to amount of suitable nesting habitat at each site.

Most nesting occurs above the high tide line, although there are some instances when a terrapin may not be savvy enough to anticipate the height of a spring tide and her nest may be flooded. If we happen to find a nest below the spring tide wrack line, we will relocate it to higher ground and hope the female's progeny won't repeat the mistake of their mother. Perhaps. it wasn't a mistake at all. Erection of bulkheads and other attempts to stabilize shorelines have created obstacles that prevent terrapins from finding suitable nesting sites.

Observation of nesting females in some locations reveals a curious behavioral pattern. Before females select a nest site, they will often sniff, taste, or probe the substrate. We have no idea what they are looking for, but this behavior is persistent and pervasive for northern terrapins. It is very common to find a female on a nesting run who has sand all over her face.

We don't know if a natal homing instinct plays a role in nest site selection. Do females return to the region where they emerged as hatchlings, or do they strike out for new nesting territory? Once females have located nesting areas, they exhibit remarkable site fidelity, a phenomenon known as philopatry, returning to the same nesting area, clutch after clutch, year after year. There are always some exceptions. In Wellfleet, we discovered a single female terrapin who thumbs her nose at the convention; she has used alternate sides of a large creek for her first and second clutches. Another exception to the rule of philopatry involved a female who produced her first clutch on June 20 and came back to the same area on July 5 to try for a second nest. She was disturbed by tourists and returned to the water, only to be discovered on July 13 on a nesting run, 10 kilometers (6 miles) farther south.

Terrapins often select sandy roads to dig nests. In Jamaica Bay Wildlife Refuge, some nesting occurs on compacted trails covered with fine gravel. This certainly makes it more difficult for females to dig their nets but they are assured sunny, nonvegetated sites. On Merritt Island, Florida, when terrapin nesting was observed in the 1970s, terrapins opted for dike roads surrounding lagoons rather than sandy dunes (Seigel, 1980b). In Wellfleet, the sandy roads and driveways on Lieutenant Island and Indian Neck are certainly well utilized. Although these nests have the potential to survive an entire summer of traffic, they are jeopardized when hatchlings begin their emergence by digging and softening the substrate. Such softened nests easily succumb to the weight of automobiles or even bicycles as they roll over the nest. If "tire track" nests are found, we will often relocate the eggs to safer areas. Care must be taken so that the order and orientation of eggs is not altered when they are relocated (fig. 3.3). Very early in development, the turtle embryo attaches to the inside wall of the egg. There will be interference with development if eggs are turned to a new position after the first few hours following their deposition in the nest. In the event that natal homing is important in the reproduction of terrapins, hatchlings that result from these nest relocations are always released at the site of their original nest.

Aside from the openness of the sites, the predilection for roadways may have something to do with the slope of the location. Although we observe terrapins nesting on steeply sloped dunes, most nests are dug on flatter surfaces. This may be partially due to the fact that nests on slopes are more difficult to dig since sand is less stable and will tend to fill in recently excavated areas. Sometimes, the only area around a marsh is steeply sloped. Such is the case in one of the locations I study each summer. Determined females head up steep hills to look for nesting sites. On several occasions, I have found nesting females because they have tumbled backward down a steep hill and landed at my feet.

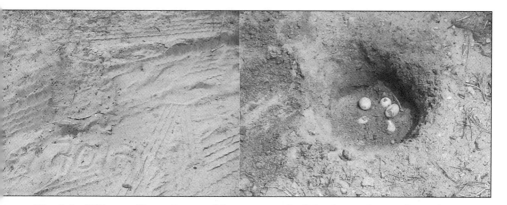

*Fig. 3.3. "Tire track" nest in a parking area on the side of a road was relocated to safer ground.*

Once the urge to nest sets in, females are resolute and unwavering. I once witnessed a young female, about eleven years old, whose two front limbs and one rear limb were whittled down to stubs. It was not clear whether she suffered from a developmental defect or a serious accident. She was gravid and had thus obviously mated. When I spotted her, she had lumbered a long distance from the marsh and was attempting to climb a steep wooded embankment in hopes of finding a suitable nesting area.

In addition to being single-minded, females on nesting forays are extremely alert. The sense of movement, the cracking of dry saltmarsh hay underfoot, the click of a camera shutter or the flight of a gull overhead will cause her to remain motionless for long periods of time. Without moving, she will blend in with the marsh vegetation or sandy substrate and will be almost impossible to discern. I have watched immobilized females for up to thirty minutes before they ventured forth to complete their nesting run or retreated back into the marsh or water. If undiscovered, they could easily be mistaken for a rock. In some instances, the patient terrapin outlasts the researcher, who must move to swat a greenhead fly or scratch a mosquito bite.

## Digging the Nest

Once a female has selected a nesting area and has made her final decision, the nesting process takes about thirty minutes. By observing twenty-eight females from the start of the process to completion, Feinberg and Burke (2003) found nesting times to vary from 13 to 47 minutes with a mean time of $24.8 \pm 6.9$

*Fig. 3.4. A female begins to dig her nest with her snout and forelimbs.*

minutes. The female terrapin sculpts her nest in an invariable sequence of events. She first smoothes and clears the area with her snout and front limbs (fig. 3.4). Then, as she alternates the scooping action of her rear limbs, she digs a small round hole approximately 4 centimeters (1.6 inches) in diameter. Still alternating her scooping, she expands the nest. Her nails help with the digging, her webbed toes with sand removal. The original hole never widens, but the nest chamber beneath slowly enlarges as she brings small loads of sand to the surface and deposits them around the nest for future use. She appears to rock from side to side as she repositions herself over the hole each time she alternates the action of her limbs. Gradually, the nest takes on a flask shape, narrow at the neck and wide at the bottom. The depth and overall measurements of nest cavities have been reported in a number of studies. Using composite data from several studies, the top of the main nest cavity varies from a mean depth of 6.0 to 10.65 centimeters (2.4 to 4.2 inches) from the surface to a mean depth of 14 to 16 centimeters (5.5 to 6.3 inches) to the bottom of the chamber (Montevecchi and Burger, 1975; Butler, 2000; Brennessel and Lewis, personal observation). Roosenburg (1994) reported the top of nests to be 5 to 17 centimeters (2 to 6.7 inches) deep (mean = 12 centimeters (4.7 inches) and the bottom to be 10.5 to 24 centimeters (4.1 to 9.5 inches) (mean = 16.5 centimeters (6.5 inches)). The nest chamber itself measures an average of 4.67 centimeters (1.8 inches) deep and 7.29 centimeters (2.9 inches) wide (Montevecchi and Burger, 1975).

Nest depth not only has a great impact on incubation temperature, it also affects the overall success of hatchling development. Very shallow nests are prone to desiccation, erosion, and high temperature stress. Deep nests are jeopardized by low temperatures and perhaps also deficits of oxygen and moisture. Burger measured the absolute success rate of hatchling development as a function of nest depth. In twenty nests that averaged 18.2 ± 2 centimeters (7.2 ± 0.8 inches), all eggs developed. In shallow nests, with mean depth of 12.5 ± 1.83 centimeters (5 ± 0.72 inches), no eggs hatched, while 5 nests having a mean depth of 14.3 ± 1.27 centimeters (5.6 ± 0.5 inches), the top eggs did not hatch. In 11 nests of mean depth 18.7 ± 1.6 centimeters (7.4 ± 0.6 inches), bottom eggs did not develop. Excavation of nests and examination of eggs that did not hatch revealed incomplete development of embryos (Burger, 1976).

When the female terrapin is satisfied with the architecture of the egg chamber, she begins to deposit her eggs (fig. 3.5). She hunkers down so that her cloacal opening is at the top of the hole. Her front limbs support her in a semierect position. She is tilted at an angle and the bottom half of her carapace may be hidden from view. Once oxytocin-like hormones have kicked in, the eggs emerge, one at a time, and are dropped haphazardly into the nest cavity. The eggs are soft, pink-tinged, symmetric ovals and have a leathery casing (plate 9). The soft texture of the eggshells make them resilient and inhibits cracking as they tumble down on top of one another. The shells will

*Fig. 3.5. Nesting female.*

dry out somewhat over the next twenty-four hours but will never become brittle like bird eggs.

Once females begin to deposit eggs, they have rarely been observed to stop midstream. It is an atypical event that prevents her from depositing the full complement of eggs into her nest. Occasionally, a female on a nesting run will drop her eggs prematurely if she is handled. Females in captivity in tanks without nesting substrate will sometimes drop eggs into the water. Without optimal nesting habitat, some females in captivity may not lay a complete clutch within the usual time period. It was once noted that captive females deposited an average of two eggs in several discrete nesting attempts spread out over two to seven weeks (Burns and Williams, 1972).

After eggs are deposited the female backfills the moist sand that she excavated, into the nest between layers of eggs, and packs the sand down with rear limbs. Her motions are the reverse of those used in the digging process. She alternates her rear limbs, scooping behind her as she backfills sand into the nest cavity. The female is not quite done when the cavity is filled. She utilizes a push-up type motion to lift herself, then thumps the nest with her plastron as she descends. She performs this up-and-down motion several times to compact the sand on top of the nest. Some observers have noticed that the females may release fluid from their cloaca over the nest, perhaps to further compact the sand or to provide moisture. If she is not frightened or rushed, the female will tidy up around the nest and kick some sand over it during her departure so that it blends in with the surrounding substrate. Sometimes the nest can be found by searching for tracks and telltale marks, especially in very sandy areas. A rather smooth area, approximately 0.25 meters (about 10 inches) in diameter, may be outlined with terrapin footprints. The thin mark of her tail may be detected over the smooth area that forms the roof over her nest (fig. 3.1). More often than not, the nest becomes an integral part of the landscape as the female terrapin leaves her eggs and future progeny to their own fate. Although some reptiles, such as crocodiles, protect their nests and attempt to assure the success of their hatchlings, turtles are completely disinterested. Once the terrapin has laid her eggs, she still has a dangerous journey back to the marsh; her return to safety appears to be a strong driving instinct after nesting. Regarding the lack of parental care, Coker (1920) remarked, "Eggs are laid in a proper place and sometimes an improper place, and so far as we know, neither parent gives thought to the welfare of its offspring or even recognizes them when they meet in passing."

## *Egg and Clutch Size*

The egg and clutch sizes published for various studies of diamondback terrapin nesting ecology display geographic variation (table 3.1). The trend is for smaller clutches with larger eggs in the southern latitudes, and larger clutches, with smaller eggs in the North. The clutch sizes of *M. t. tequesta* in central Florida (Seigel, 1980b) average 6.7 eggs, similar to those reported for *M. t. centrata* in northeast Florida (Butler, 2000; Butler, et al., 2004). In Jamaica Bay, New York, the mean clutch size increases to 10.9, with a wide range of three to eighteen eggs per clutch (Feinberg and Burke, 2003). Wellfleet, Massachusetts, terrapins have average clutches of twelve eggs, with a range of four to twenty-two eggs per nest (Don Lewis, personal communication). The exception to this latitudinal correlation is the somewhat larger clutch and eggs size of Chesapeake Bay terrapins (average 12.29 eggs with a mass of 9.87 grams [0.35 ounces]) (Roosenburg and Dunham, 1997). In some studies conducted within local populations, clutch size appears to correlate with the size of the female. In general, the tendency is that the larger the female (as measured by mean plastron length), the larger is the clutch (Montevecchi and Burger, 1975; Seigel, 1980a; 1980b; Roosenburg and Dunham, 1997). Depending on climate, it may be more useful for terrapins to produce multiple small clutches, which occurs in the South, or fewer, larger clutches, which may be more typical of northern regions with a more contracted nesting season. Terrapins at the northern fringe for the species may be putting "all their eggs into one basket."

One might be tempted to predict that within a terrapin colony, smaller clutches would contain larger eggs or that larger females would have larger eggs. But studies have shown that within a colony, there is no correlation between clutch size and egg size. There is also a lack of correlation between the size of a female and the size of her eggs (Montevecchi and Burger, 1975; Roosenburg and Dunham, 1997; Seigel, 1980b). In general, terrapins have a large variation in eggs size among clutches but little variation within clutches. In one study, egg size tended to decrease with clutches laid later in the season and no differences were found in the size of eggs deposited in different topographic regions of the nesting area (Montevecchi and Burger, 1975).

The trade-off between larger clutch size and larger egg size is not completely clear. Egg size may be an important contributor to hatchling survivorship. Larger eggs, with more food reserves, usually produce larger hatchlings. This may be an important strategy in habitats in which hatchlings will be competing for resources. In contrast, production of more offspring may be a

**Table 3.1.** Summary of Diamondback Terrapin Nesting Studies

| Study site | Nesting season | Time of day | Weather | Substrate |
|---|---|---|---|---|
| Louisiana | — | — | — | — |
| Northeast Florida | Late April–late July, 78 days | Diurnal and nocturnal | — | — |
| Central Florida, east coast | Late April–early July, 52–57 days | Diurnal | Clear skies | Dike roads, compacted sand |
| Beaufort, N.C., natural nests | — | — | — | — |
| Beaufort, N.C., turtle farm | 6 May–31 July, 80–90 days | — | — | Artificial sand pans |
| Patuxent River, Maryland | 1 June–30 July, 60 days | Peaks at 1100 to 300 h; observed round the clock | Sunny | Narrow sandy beaches |
| Cape May, N.J. | Early June–mid July, 41 days | — | — | Road embankments |
| Brigantine National Wildlife Refuge, N.J. (now Edwin B. Forsythe National Wildlife Refuge) | 34–44 days | Diurnal | 25–75% cloud cover | — |
| Jamaica Bay Wildlife Refuge, N.Y. | 3 June–13 July, 51 days in 1999; 9 June–4 August, 57 days in 2000 | Diurnal 0930 to 2115 h | — | Partially vegetated sandy areas and gravel trails |
| Barrington, R.I. | 10 June–13 July, 34 days | Diurnal mostly in morning | — | Non-vegetated, sandy areas |
| Barnstable, Mass. | — | 45% nocturnal | — | 50% on vegetated dunes; 50% on open dunes |
| Wellfleet, Mass. | Mid-June–late July, 23–42 days | Mostly diurnal; some nocturnal | Clear to partly cloudy skies | Sandy to partially vegetated dunes |

| Average nesting trek | Mean clutch size | Mean egg mass (grams) | Number of clutches/ terrapin/year | Reference |
|---|---|---|---|---|
| — | 8.5 | — | 1 observed | Burns and Williams (1972) |
| — | 6.7 ± 1.4 | — | Up to 3 | Butler (2000); Butler et al. (2004) |
| Short; nesting areas near water | — | — | Up to 3 | Seigel (1980b) |
| — | 5.29 | — | — | Coker (1906) |
| — | 8 | — | Up to 5 | Hildebrand (1932) |
| <10 m | 12.29 ± 0.13 | 9.87 (0.35 oz) | Up to 3 | Roosenburg (1994), Roosenburg and Dunham (1997) |
| — | 8–12 | — | — | Wood and Herlands (1995) |
| <100 m | 9.76 ± 2.61 | 7.7 (0.27 oz), range = 5–11 (0.18–0.39 oz) | — | Montevecchi and Burger (1975), Burger and Montevecchi (1975) |
| <100 m* | 10.9 ± 3.5 range = 3–18 | — | Up to 3 | Feinburg and Burke (2003) |
| Approx. 10–100 m, some up to 500 m. | 11.9 | — | Up to 2 | Goodwin (1994) |
| Long treks up to 1600 m | — | — | — | Auger and Giovannone (1979) |
| Long treks are common | 12 range = 4–22 | 7.75 (0.27 oz) range = 4.5–11 (0.16–0.39 oz) | Up to 2 | Lewis (personal communication) |

*personal observation

better strategy when hatchling mortality is high. Bitter winter temperatures that decrease soil temperatures below zero for prolonged periods may result in considerable hatchling mortality. Draud has shown a 50 percent temperature-induced mortality of terrapin hatchlings in Oyster Bay during the relatively cold winter of 2003 to 2004 (Draud , Zimnavoda, King, and Bossert, 2004). Mortality due to low temperatures would not have such a devastating impact in southern latitudes. Another possible reason for larger clutches in northern colonies may be the result of seasonal limitations in producing two clutches. Northern terrapins that produce a single clutch may be able to produce the same number of offspring per season as their Southern cousins who have the time to produce two clutches.

Roosenburg looked at the possible correlation between egg size and nesting location. Since egg mass correlates with hatchling size, he wondered where the larger eggs are laid. In the case of females, larger eggs that produce larger hatchlings would create a scenario in which females may mature several years before those that come from smaller eggs. (This relationship does not hold for males because both large and small male hatchlings reach maturation size in approximately the same time frame.) Roosenburg proposed a model for nesting in which placement of larger eggs under conditions that produce females would benefit the species. From this reasoning, Roosenburg hypothesized that larger eggs should be placed in open sites with warmer incubation temperatures than smaller eggs. Data collected at his field site in the Patuxent River, support this idea (Roosenburg, 1996).

## Incubation Temperature and Development

For the entire range, the average incubation period for terrapin eggs is sixty to ninety days. Egg development is a function of environment, and hence soil temperature and moisture. The warmer the environment, the faster will development progress. The relatively shorter time required for development of a terrapin hatchling in Southern locations where spring temperatures are relatively warm allows multiple clutches per female per year, but cold spring and early summer temperatures may sometimes limit Northern females to one clutch per year. Auger and Giovannone (1979) reported an average incubation period of 108 days (range = 87 to 146) on Cape Cod. A more recent Cape Cod study (Lewis and Prescott, personal communication), conducted from 2000 to 2002, indicates a mean incubation period of 81 days (range 59 to 116 days), while Burger (1977) reported an average of 75 days (range 61 to 104) in New Jersey.

Temperature also has a profound effect on gonadogenesis (development of gonads) and thus on the resulting sex of each hatchling. As mentioned in chapter 1, diamondback terrapins lack X and Y or sex-determining chromosomes. Sex is strongly influenced by temperature. The phenomenon of temperature-dependent sex determination (TSD) is prevalent in most turtles. The benefit of TSD for a species is not completely clear. Nonetheless, the lack of sex chromosomes that guide the genotypic development of males and females has not hindered reptiles from producing progeny of both sexes.

In the laboratory, incubation temperatures below 28°C (82°F) produce male terrapins, while temperatures above 30°C (86°F) produce females. Intermediate temperatures result in a mixture of males and females (Jeyasuria and Place, 1997). In natural nests, temperature does not mimic controlled laboratory conditions. Furthermore, in some laboratory studies, the effect of temperature is not as nicely delineated. For example, in a laboratory setting with incubation temperatures set at 26, 28, 30, 32, and 34°C (79, 82, 86, 89, and 93°F) the percentage of male hatchlings was 100, 93.3, 11.1, 0.0, and 7.7, respectively (Giambanco, 2002). There is no clear explanation for the development of one male from thirteen eggs incubated at the highest temperature. For each turtle, TSD is believed to be an all-or-none phenomeno; hermaphrodites have rarely been observed.

In nature, incubation temperature is not constant; it fluctuates every day and also as a function of the time of year. In the early 1970s, Burger (1976) conducted studies of temperatures in four natural nests on Little Beach Island in New Jersey. Two of the nests were on south-facing slopes; the other two were on north-facing slopes. One of her first observations was a diel (daily) temperature variation of 2 to 12°C (approximately 4 to 22°F). In Maryland, a diel variation in soil temperature of as much as 10°C (19°F) was also observed (Jeyasuria et al., 1994; Roosenburg, 1996); while on Cape Cod we have recorded temperature variations of 2 to 7°C (about 4 to 13°F) per day (fig. 3.6). In the New Jersey study, daily low temperatures occurred at 0600 h and daily highs at 1500 h (Burger, 1976). In Maryland, the daily low temperature also occurred at 0600 h, while the daily high was measured between 1300 and 1400 h. On Cape Cod, daily lows also occur at about 0600 h and highs at 1600 to 1800 h (fig. 3.6). In New Jersey, the mean low temperature ranged from 19-24°C (66 to 75°F) and the high ranged from 23°C to 31°C (73 to 88°F). This compares to Cape Cod where we recorded mean nest temperatures from a low of 24°C (75°F) to a high of 31.5°C (89°F) in 2003. Nest high and low temperatures as well as diel temperature variations will be different from year to year,

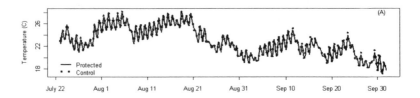

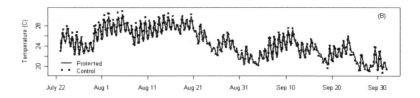

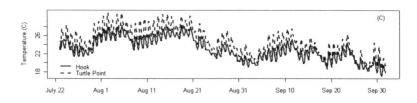

*Fig. 3.6. Temperature recorders trace diel and seasonal variation in the tempera-ture of soil in a terrapin nesting area. A and B: Solid lines are soil temperatures; dotted lines are soil temperatures under an adjacent predator excluder device. Site A was Turtle Point, a sunny, south-facing slope; site B was the Hook, a north-facing slope. There were no differences in temperatures of soil under predator excluders compared to adjoining soil. In graph C, the soil temperatures in the two sites are compared. This graph demonstrates that there are significant differences in soil temperatures at nesting sites.*

depending on the weather conditions. Another parameter affecting incuba-tion temperature is reflected in the difference between north- and south-fac-ing slopes. North-facing slopes sustain slightly cooler nest temperatures, and hence slightly longer incubation periods (79 ± 3 days compared to 71 ± 3 days) (Burger, 1976). We have also found a significant mean temperature dif-ference of 1 to 2°C (1.8 to 3.6°F) per day when north- and south-facing nests are compared throughout a nesting season (Brennessel and Lewis, personal observation).

One of the major differences in Burger's study in the 1970s compared to our observations in 2003 is the detection of slightly higher temperatures in nests during the last three weeks of incubation compared to surrounding soil. Burger attributed this difference to metabolic heat. As yet, we have found no evidence for the production of metabolic heat in nests on Cape Cod.

Nest depth has an impact on temperature of incubation. Placement of an egg in the nest may also influence development time and sex of the resulting hatchling. Temperatures at the top of nests are several degrees higher than temperatures near the bottom of the nest. There is an approximately 2°C (3.6°F) difference in positional nest temperature when the top and bottom of a nest is compared. Furthermore, the top of the nest experiences a greater variation in diel temperature than the bottom, suggesting an insulating effect in the deeper regions of the nest (plate 10). Superimposed on the diel temperature variance in these shallow terrapin nests, seasonal and occasional temperature variations are also important to consider. During some seasons, a protracted period of cool or hot weather may have a role in determining the turtle phenotype. The time that the nest is laid within a season will also be a factor. Early spring clutches will be subjected to gradually increasing temperatures through June and July; late-season clutches may experience progressively decreasing temperatures at the end of summer or beginning of fall (Shine, 2004).

The temperature of nests throughout a season can be important in determining the time it takes for hatchlings to develop, but the sequence of temperature changes in the nest will be more important for TSD. In turtles, there is a narrow window of time during development, which usually occurs in the middle trimester, in which temperature must reach a critical threshold to trigger sexual differentiation. Prior to this critical time period, the embryonic turtle is neither male nor female; its gonads are said to be bipotential or indifferent (Porter, 1972; Pieau and Dorizzi, 2004).

The diel and seasonal variations in nest temperature depicted in fig. 3.6 and plate 10 were obtained by placing a temperature probe in a protected terrapin nest. The probe recorded temperature every thirty minutes during the entire course of the incubation period. The nest depicted in plate 10 was a late-season nest, laid on July 20, 2003, on a sunny, south-facing slope. Fifteen very small hatchlings (average weight 4.4 grams [0.16 ounces], range = 3.5 to 5 grams [0.12 to 0.18 ounces]) emerged from the nest, only eight of which survived. All surviving hatchlings were females. Even thought the nest was laid late in the nesting season, in mid July, soil nest temperatures at the top of the nest reached 30°C (86°F) or higher (the temperature that produces females) on twenty days during the early incubation period. It is interesting to note that nest temperatures were maintained at high levels for short periods of time. On some days, high temperatures were reached for one to two hours; on other days, temperatures reached over 30°C (86°F) for up to six hours. Throughout the incubation period, high temperatures were reached during the afternoon, from 1300h-1600h. This nest was coolest from about 0600 h to 0800 h. In this nest that produced females, the average temperature was 24°C (75.2°F); the temperature varied from a low of 16°C (60.8°F) to a high of 35°C (95°F), with warmer temperatures during the first half of the summer, overlapping with the critical period for TSD.

In another study of the effects of temperature on terrapin sex determination, Giambanco (2002) correlated mean temperature, incubation period, and sex of hatchlings to solar exposure in three nests. One nest had no direct sunlight, one was shaded two to four hours each day, and one was in full sunlight. The nests hatched within eight days of each other. However, the nest without direct sun had slightly cooler mean, minimal, and maximal temperatures than the other two nests and produced over 90 percent males. The partially shaded nest produced a mixture of males and females. The nest in full sun had the highest minimal, maximal, and mean temperatures but produced only three viable hatchlings; all were female.

If sex of diamondback terrapins is influenced by temperature within a critical time period, we might ask how temperature dictates the outcome of sexual development. The explanation may hinge on biochemical events that affect the expression of specific genes. These events may be similar whether temperature or sex chromosomes provide the biochemical switch that initiates the sex-determining program. A likely candidate gene that may be key in sexual determination in vertebrates contains the code for an enzyme known as P450 aromatase. This enzyme catalyzes the formation of "female" steroid hormones (estrogens) from androgens in two areas of the body: the brain and gonads. Estrogens such as estradiol are capable of switching the sex of turtles incubated at male-producing temperatures so that they develop as females. Thus, we can consider the early embryonic turtle gonad to be sexually "indifferent." Estrogens will inhibit the formation of testes and stimulate the formation of ovaries. The P450 aromatase enzyme is much higher in gonads of diamondback terrapins incubated at female-producing temperatures than at male-producing temperatures (Jeyasuria et al., 1994; Jeyasuria and Place, 1997; Place et al., 2001).

The explanation of TSD in diamondback terrapins is not as simple as a temperature-induced expression of P450 aromatase. Timing of embryonic estrogen synthesis in brain and gonads and amount of estrogen produced are other critical factors. Development of embryos occurs by a very specific sequence of events. The pattern in which these events unfold is key to TSD. Diamondback terrapin embryonic development has not been studied in detail, so most of our baseline knowledge of terrapin developmental stages comes from studies of snapping turtle (*Chelydra serpentina*) eggs, incubated in the laboratory at 20°C (68°F) (Porter, 1972). Under such conditions, investigators have categorized a presomite period (stages 0 to 3), in which the embryo is a tiny blob that can be described microscopically; a somite period (stages 4 to 10), in which differentiation of major body sections occurs—the head region can be distinguished from the tail region and placement of major organs can be discerned; and the limb period (stages 11 to 26), in which development is completed as the turtle shell is formed. The middle embryonic stages, 12 to 14, appear to be the critical times during development for TSD. The temperature-sensitive period coincides with the same developmental stages as the estrogen-sensitive period. This period is the developmental window in which administration of estrogen can shift development from male to female at male-producing temperatures. Chemical-induced inhibition of estrogen production or action during this period will produce males at female-producing temperatures. In snapping turtles, this stage occurs at approximately 30 to 42 days at 20°C (68°F), but proportionally faster at higher temperatures of incubation.

Place et al. (2001) used the basic staging hallmarks described for snapping turtles in their study of the expression of P450 aromatase in diamondback terrapin embryos incubated at male-determining temperature (26.5°C; 79.7°F) and female-determining temperature (30.5°C; 86.9°F). The temperature-sensitive period when sex is determined in diamondback terrapins incubated at 26.5°C (79.7°F) was between days 20 to 40, and for incubation at 30.5°C (86.0°F) it occurred between days 15 to 30. Place et al. found that differences in aromatase expression and production prior to embryonic stage 18 are the most critical for TSD. After this stage, it is very difficult to manipulate the sex of terrapin hatchlings by changing temperature or by using agents that inhibit aromatase activity. Place et al. (2001) believed that the interplay of aromatase activity between brain and gonads is critical for the sexual development in diamondback terrapins. In contrast, a review of TSD in other turtles suggests that the mechanism for TSD is confined to the gonads (Pieau and Dorizzi, 2004).

The expression of P450 aromatase and the subsequent production of estrogens are clearly important in TSD, but there are missing links in our knowledge of the detailed temperature-dependent series of biochemical events that dictate sexual development. Much of the differentiation process involves a complex network of genetic regulatory mechanisms that play out at the molecular level and may be similar to those that control genotypic (determined by sex chromosomes) sex determination.

Given the effect of temperature on sex determination in terrapins, we can explore whether terrapins specifically manipulate the sex of offspring by selecting certain nesting sites over others. When Doody et al. (2003) asked this question about the pig-nosed turtle (*Carettochelys insculpta*), a turtle with TSD that inhabits rivers and billabongs in northern Australia and New Guinea, they found that the turtles did not favor beaches that were warmer or cooler than those that were generally available. The researchers concluded that Australian pig-nosed turtles do not manipulate sex of their offspring by beach selection.

Roosenburg (1996) asked the question about nest site selection in diamondback terrapins in a more complex scenario. He introduced the variable of egg size and concluded that terrapins select warmer nesting sites when their clutches contain larger eggs. This strategy would give a head start to larger female hatchlings that take longer to reach sexual maturity than male hatchlings.

Incubation temperature may also affect other aspects of reproductive success and hatchling development. In other reptiles, incubation temperature has been shown to have an impact on hatchling traits such as size and locomotor activity. To date, no one has reported on the effects of incubation temperature on hatchling size and other traits that may affect the ability of the hatchling to survive after emergence or reproduce in later years.

The global question about TSD is: Why does it occur at all? Even more perplexing is the fact that as common as TSD is in reptiles in general and turtles in particular, there are some emydid turtle species such as *Clemmys insculpta* and some *Apalone* (soft-shelled turtles) that have abandoned this strategy and rely on sex chromosomes to determine whether an egg develops as a male or female. The chromosomal strategy is referred to as genetic sex determination (GSD). There is no obvious reason why TSD, in contrast to GSD, would be an adaptive advantage for terrapins. With GSD, due to chance, there would be an equal probability of producing males or females. With TSD, however, the location, placement, and depth of the nest become major factors in the resulting sex ratios of offspring. Instead of chromosomes, the nesting behavior of the mother turtle becomes the key component of sex determination.

It could be postulated that GSD may be an advantage for turtle species at the extremes of their range. For example, it would not be beneficial for Southern subspecies if all Southern nests produced primarily females. It will also not be beneficial if all Northern nests produced mostly males. GSD would eliminate the possibility of such skewed sex ratios. However, even with TSD, there is no evidence for such skewed ratios among hatchlings of any turtle species. Furthermore, there is no geographic variation in expression of TSD versus GSD among turtles.

From a study of the pattern of TSD utilization in twenty-two turtle species, a biological association between nest temperature and adult sexual dimorphism has been suggested (Ewert and Nelson, 2002). The sex that predominates at cooler temperatures is usually the smaller turtle. This is certainly the case for diamondback terrapins. Perhaps incubation temperatures produce physiological differences that optimize the fitness of adults. Appealing as this theory is, it does not account for those instances of TSD in turtle species in which males and females are approximately the same size.

Another possible adaptation that has been proposed to explain TSD is sib-avoidance. Since TSD often produces same-sex clutches, inbreeding can be prevented. GSD would produce approximately equal numbers of males and females that could potentially interbreed. This explanation doesn't quite fit the diamondback terrapin reproductive strategy. Diamondback terrapins become sexually mature many years after they disperse from their nests. They do not form pair bonds, nor are they expected to find the same mates each season. The possibility of diamondback terrapins producing full siblings of the opposite sex from different clutches may be nonexistent. The lifetime yield of a mating between any individual female and a specific male may be only a single clutch.

Could TSD be a group structure adaptation (Ewert and Nelson, 2002)? Perhaps the strategy keeps the ratio of male to female diamondbacks optimal for the population. TSD has the potential to produce more females than males, which may offset the earlier maturation of males and hence the tendency of populations to exhibit a male bias. We don't know the ratio of female to male hatchlings in most natural settings, but from the available studies it appears that many populations stabilize with an adult sex ratio that favors females.

## Nesting Outcome

There are various estimates of the outcome of nesting efforts of female terrapins in the wild. Sometimes an egg or two in a clutch will fail to develop. Even if development occurs, some eggs will not produce viable hatchlings. The female terrapin never knows the outcome of her nesting foray. Should her nest be too shallow or too deep, hatchlings will not develop properly. Natural forces such as wind erosion and storm surges that inundate or wash away nests can certainly take their toll on eggs deposited in vulnerable locations. There is no way for the individual female terrapin to learn from her mistakes. Each year, we find a terrapin nest, most likely laid by the same clueless female, in an area that is subject to tidal inundation during spring tides. Perhaps the eggs can survive several floodings, but probably not a whole season of spring tides that wash over and may even erode the nest twice a month. We can relocate these eggs to drier ground and release them in their natal location after they hatch, but if natal homing is at work, the female's hatchlings might return to the same vulnerable nesting area year after year when they are sexually mature.

Aside from environmental destruction, other factors may influence the outcome of successful hatching. Animal predators are the most important contributor to nest failure. In many locations throughout the range of the diamondback terrapin, the raccoon, *Procyon lotor*, is a major predator of terrapin eggs, especially within the first twenty-four hours after the eggs are laid. Most likely a transient indicator, such as scent, is used by raccoons to locate nests. Raccoons also display keen interest in areas that have been recently excavated. Nest depredation also occurs after more than twenty-four hours of egg deposition, but at a much lower frequency. In some instances, researchers mask the scent of a freshly laid nest by providing an alternate scent (e.g., human urine) until the nest survives its first twenty-four hours or until it can be effectively protected in another manner.

Raccoons are categorized as a subsidized predator because their numbers increase in areas where people live. They are attracted to locations where they have access to trash and other easy sources of food and they can prosper. These subsidized predators have greatly changed the outcome of nesting success for diamondback terrapins in Jamaica Bay Wildlife Refuge (JBWR). There were no signs of raccoons or nest depredation in the refuge in the late 1970s and early 1980s when Bob Cook first studied diamondback terrapins in this national park. Raccoons were introduced and/or found their way into the refuge in the mid 1980s and became fairly common in the 1990s. During a study conducted in 1998 and 1999, raccoons depredated 92.2 percent of terra-

*Fig. 3.7. Field marks of raccoon predation.*

pin nests on Ruler's Bar Hassock, the prime terrapin nesting area in the refuge (Feinburg and Burke, 2003). During that time, raccoons were seen in the nesting area mostly at night. During my visit in 2004, several raccoons were spotted in the nesting area, bold as can be, in the middle of the day. One was eating terrapin eggs as the female deposited them in the nest (plate 7; described in chapter 2). Only the continued presence of researchers, volunteers, and refuge visitors prevented massive terrapin egg carnage during daylight hours.

Raccoons usually leave a pile of shell fragments close to the nest, and indeed, the presence of shells is often a field mark of a terrapin nest (fig. 3.7). At JBWR, raccoons also leave shells in conspicuous piles near the depredated nests. However, Feinburg and Burke (2003) found a curious behavior during the latter part of the nesting season: It appeared that raccoons sometimes consume entire eggs, including shells. No eggshells were found near certain marked nests that were depredated. Furthermore, raccoon scat (feces) contained large amounts of terrapin eggshells. The biological basis for the change in raccoon feeding behavior is not known and has not been extensively observed in other terrapin nesting areas. Butler (2000) and Burger (1977) reported some cases in which raccoons consumed entire eggs, but in general the shells are left behind. Shells may also be missing from nests if eggs are carried off by gulls to feed their chicks (Burger, 1977).

Raccoons are also the main terrapin nest predator at Sandy Neck, Massachusetts (Auger and Giovannone, 1979), and on Little Beach Island, New Jersey (Burger, 1976). At some sites, rates of predation vary from year to year for reasons that cannot be easily explained. On Little Beach Island, New Jersey, in 1973, at least some eggs developed in 84 percent of nests. Thirty-nine percent of all eggs produced hatchlings. In 1974, some eggs developed in only 25 percent of nests. Eighteen percent of all eggs produced hatchlings. These differences in hatchling success rates were entirely due to predation.

In Wellfleet, Massachusetts, different nesting areas have suffered different degrees of nest depredation by raccoons. In more developed locations, where suitable nesting areas are at a premium, over 90 percent of nests are consistently lost to raccoons. In less populated nesting locations, such as those under conservation protection by the Massachusetts Audubon Society, predation occurs at a lower frequency. On Cape Cod, raccoon predation occurs mostly, if not exclusively, at night.

Other animal predators of diamondback terrapin eggs include red fox (*Vulpes fulva*), American crow (*Corvus brachyrhynchus*), laughing gulls (*Larus atricilla*) in New Jersey (Burger, 1977), fish crows (*Corvus ossifragus*) in northeast Florida (Butler, 2000), red fox and northern river otter (*Lutra canadensis*) in Maryland (Roosenburg, 1994), and ghost crabs (*Ocypode quadrata*) in Florida (Arndt, 1991, 1994; Butler et al., 2004). The avian predators may be more successful at the time of egg laying (Burger, 1977). Butler (2000) observed clever crows following gravid females to nesting areas and patiently observing researchers as they processed nests.

Relatively recent reports indicate that other animal predators may be making an impact on diamondback terrapin reproduction: rats! On mangrove islands in Florida Bay, black rats have been a recent introduction and may be responsible for predation on nests (Mealy et al., 2004). A similar trend was seen during terrapin farming operations in the early twentieth century in North Carolina where rats ate both eggs and hatchlings (Hildebrand and Hatsel, 1926). In Oyster Bay, New York, the Norway rat apparently found a new food source in diamondback terrapin hatchlings (Draud, Bossert, and Zimnavoda, 2004). On nesting beaches, rats are a nocturnal predator. They have not been observed to disturb nests, but their teeth marks are evident on newly hatched terrapins (further described in chapter 4).

The site of the nest may affect the outcome of reproductive success. Burger (1977) found that nests near vegetation were more prone to predation by mammals, and those in open sandy areas to predation by birds. She also found that nests in high-density nesting areas, within 1 meter of each other, were preyed upon at a higher rate than nests that were dispersed. I have often observed a large number of predator digs in historically high-density nesting areas, sometimes up to two weeks before nesting begins. Predators such as raccoons have cued in to high-density nesting areas and may not need visual or olfactory cues to locate nest. Random digging in these areas will have a high probability of unearthing tasty terrapin eggs.

Although most incidents of nest predation by animal predators occur within the first day or two after eggs are laid, another spike of destruction affects nests as hatchlings are ready to emerge. On Little Beach Island in the 1970s, there was actually more nest depredation during hatching than nesting, occurring before all the hatchling turtles emerged from the nest (Burger, 1977). In Wellfleet, during September and October, we often locate nests that we have missed during the egg-laying season when we see signs of predation such as piles of broken shells and relatively large, predator-initiated digs in the sand. Perhaps the emergence of the first few hatchlings alerts predators to the nest location, either by scent or by visual cues.

Not only do the local fauna take a bite out of the future generations of terrapins, the flora can also destroy terrapin eggs. Roots from plants such as dune grass (*Ammophila breviligulata*) infiltrate eggs and halt their development (fig. 3.8). In some cases, the entire nest is invaded by roots even before development has progressed. Lazell and Auger (1981) were surprised and impressed by how rapidly roots can grow and infiltrate nests. The roots penetrate and pack the eggs. In other cases, development is almost complete before hatchlings become strangled by plant roots. We have observed dead, almost fully

*Fig. 3.8. Eggs that have been infiltrated with plant roots.*

developed hatchlings with ligature marks inflicted by plant roots.

It is sometimes easy to predict the location of nests that roots or rootlets will infiltrate. In vegetated areas, it is nearly impossible for diamondback terrapins to find a clear sandy patch to nest in. Therefore, terrapins will often nest in eroded areas, formed by humans and nonhuman animals as they trek through marsh uplands. If the eroded areas fill in with plants during the summer, terrapin eggs will be in jeopardy. In most cases, during the early part of the nesting season there are no clues for terrapins to predict the future location of luxuriant vegetation. By the end of August, a formerly bare dune may be covered in dense dune grass. I have protected nests from animal predators in late June only to have plants destroy the nests in July and August. Auger and Giovannone (1979) questioned whether the plant roots actively seek the moisture and nutrients contained in the egg, or whether root infiltration represents a chance event. No one has provided a satisfactory answer to this query.

Insects are an unlikely but significant predator of diamondback terrapin nests, affecting both eggs and hatchlings. On Sandy Neck, Barnstable, Massachusetts, approximately 35 percent of nests laid in 1978 were victims of at least partial maggot infestation. The Systematic Entomology Department of Agriculture identified the maggots from Barnstable, Massachusetts, nests as larval forms of a member of the flesh fly family, *Sarcophagidae*. In Wellfleet, partial or entire nests full of small, fully formed terrapins suffer mortality when the hatchlings are infected by maggots. The route of entry in many cases appears to be via the yolk sac. Various species of ants have also been observed in nests, but in some cases it is not clear if they feeding on crushed eggs or dead hatchlings or if they are responsible for egg or hatchling fatality.

Predation on diamondback terrapin nests seems to be increasing at a dramatic rate. Whether this is the result of more comprehensive examination and reporting or actual decreases in reproductive success needs to be clearly defined.

Whether it is a raccoon, plant, or insect, predation on terrapin nests has been a facet in the course of this turtle's natural history. As long as a sufficient number of hatchlings survive to reproductive age it is unlikely that the population will suffer. But the question becomes: What constitutes a sufficient percentage of hatchling survival to recruit turtles into the reproducing component of population? If there are additional stresses on the population, including those on juveniles and adults, it may not be possible for terrapins to sustain high nest depredation rates and still maintain stable populations.

# Chapter 4

# The Lost Years

ON HANDS AND KNEES, Matt Draud and his students from C. W. Post College, a branch of Long Island University, are parting the grass and shrubbery and scouring the marsh at a town beach in Oyster Bay, New York. They are in pursuit of diamondback terrapin hatchlings. Each orange flag designates the location where a small turtle has been found. The area looks like an obstacle course for tiny creatures; it is studded with dozens of orange flags. The location of all this activity is a patch of vegetation about a meter wide that consists mostly of *Spartina patens*. This is one of the prime hiding and foraging habitats for terrapin hatchlings in Oyster Bay.

Draud is one of several scientists trying to solve the mystery of the "lost years"—where and how diamondback terrapins spend their first years. Although newly hatched terrapins are relatively easy to find if we already know the nesting areas within a colony and locations of individual nests, it is much more challenging to find young and juvenile terrapins. In contrast to the wealth of information on nesting activity of females, the period between hatching and maturity is the black hole in terrapin life history.

If we base our assumptions about brackish water terrapins on observations of marine turtles, we would expect hatchling diamondback terrapins to make a beeline for the water to get out of harm's way. As miniature sea turtles emerge from their underground nest chambers, predators lurk in anticipation. Birds circle overhead and are ready to clutch baby sea turtles the moment they emerge from the nest and scramble toward the water. Predators are not limited to the skies. Baby sea turtles make tasty morsels for a variety of marine life. Once sea turtle hatchlings safely complete their scramble to the ocean, the further travels of hatchlings are not well understood. Although there has been some progress in tracking their migrations and foraging patterns, for the

*Fig. 4.1. Emergence hole provides escape from nest.*

most part they disappear for a few years and are seen again when they are much larger. The term "lost years" was first coined to reflect our lack of knowledge about this period in marine turtle life history.

Parallel to development of marine turtles, the diamondback terrapin also has some "lost years." It is this gap that Draud and others hope to fill as they rake through the vegetation in diamondback terrapin nesting areas, looking for signs of tiny terrapins.

## Emergence

Nature equips turtle hatchlings with a tool that is also found in newly hatching birds, the egg tooth (plate 11). This structure is not actually a tooth, although it looks like one. The egg tooth is a tiny protuberance at the tip of the nose that functions like a box cutter. When a terrapin is about to hatch, it uses its egg tooth to slice an opening in the eggshell. This produces an exit point from which a tiny limb may emerge to lengthen the opening and eventually free the hatchling from its shelled home (plate 12). The process of breaking an exit hole in the eggshell is known as pipping. Once a hatchling has pipped, it may remain in the nest, and even within the remnants of the shell, for a considerable period of time, sometimes as long as the following spring. During the time the hatchling remains in the nest, the yolk sac, containing important nutrients, is resorbed and used for energy. No longer needed once an exit hole in the eggshell is produced, the egg tooth is also resorbed.

Hatchling diamondback terrapins from a single clutch may not all emerge in a synchronous manner. They sometimes stagger out, a few at a time, over a period of a few days to almost two weeks. Some hatchlings will not emerge in this time frame. In some instances, a fraction of the hatchlings will emerge while other nestmates will opt to get through their first winter hunkered down in the nest.

The journey from the nest initially involves an excavation project. The sandy or gravel soil over the nest usually becomes compacted during the summer months and may be held together with plant roots. Some hatchlings may break out on their own, but the concerted efforts of the hatchlings help to loosen the nest cover so that emergence is facilitated. If we brush sand from the top of the nest to assess hatching progress, the nest that is ready to erupt with hatchlings will be bubbling with underground movement. As the first few hatchlings emerge from the confines of the nest they produce one or more emergence holes, small tunnels having the diameter of a quarter, about the width of a hatchling carapace (fig. 4.1). These emergence holes can sometimes be mistaken for the entrance to a crab burrow although crab holes are typically found in wetter areas of the marsh. When hatchlings first push their heads out of the sand, their eyes are closed. They have sometimes been observed in this position for two to four hours during which their eyes open and their heads move in the direction of the sun (plate 13). This emergence behavior may be responsible for setting an internal biological clock and may function as part of a solar orientation mechanism (Auger and Giovannone, 1979).

The sight or scent of emerging hatchlings will attract predators. Many terrapin colonies experience a second wave of nest depredation in the late summer and fall, when hatchlings begin to scramble from their nests. Among the predators of emerging hatchlings are raccoons, birds, rats, and ghost crabs. Raccoons and birds will make off with whole hatchlings, while rats and ghost crabs nibble away on soft parts and leave the shell and uneaten parts in the marsh. Although rats have not been observed to depredate eggs or hatchlings within a nest, they will attack hatchlings once they have emerged and kill them by evisceration through the carapace or plastron. Over a three-year period, Matt Draud and his students observed nocturnal predation by Norway rats during hatchling emergence in August and September. The rats know to come back for more. They return in April to prey upon yearling terrapins as they emerge from hibernation (Draud and Bossert, 2002; Draud, Bossert, and Zimnovoda, 2004).

The hatchlings that remain in the nest the longest may be most vulnerable. Burger (1976) reported a study in which hatching occurred over a one to four day period, with individual hatchlings emerging from the nest as many as eleven days apart. All hatchlings emerged during the warmest period of the day, most commonly between 1200 and 1700 h. She observed that earlier emergence reduced an individual turtle's chance of being found by a predator. Some hatchlings will linger. They opt to remain in the nest over the entire fall and winter and make their break for the outside world the following spring, perhaps when environmental conditions become more favorable.

The pattern of prolonged hatchling emergence may affect the sex ratio of hatchlings that eventually disperse into the marsh. If females are produced in the upper area of the nest, where temperatures are usually higher during the incubation period, and if they are the first to emerge, they may have the greatest chance of short-term survival. Males, found in deeper regions of the nest, may be at the mercy of predators that have cued in to the nest location by the activity or scent of hatchlings that are first to emerge.

## The Hatchling's Journey

In some field studies, it has been estimated that only one out of five hatchlings may complete the journey from the nest to a safe location within the marsh (Auger and Giovannone, 1979). Many will succumb to predation, overheating, or desiccation. In some nesting areas, the nearest body of water is within 10 meters (11 yards) of the nest. On Cape Cod, a tiny hatchling may be required to travel a kilometer (0.62 mile) or more to reach the nearest body of water or the marsh (plate 14). How does the emerging hatchling survive the gauntlet of awaiting predators? If the hatchling is making its way to the water, what mechanisms guide it? How does the hatchling navigate from the uplands to a marsh that may not be visible because of dense vegetation, sand dunes, roadways, or other obstructions? Does the hatchling have an internal orientation program. and if so, what are the cues that it uses? These are questions that have not been successfully answered in terrapin studies, although some progress has been made to address these aspects of emergence in sea turtles. Many sea turtle hatchlings emerge at night to avoid avian predators, and they use moonlight and the horizon as cues to guide them to the water. House and street lights near sea turtle nesting beaches have been known to distract hatchling sea turtles and send them in the wrong direction. In contrast, terrapins emerge during daylight hours and must use different cues to guide them out of the nest and into safe locations.

Several studies have addressed the behavior and travels of hatchling terrapins after they have left the nest. In the mid 1970s, field and laboratory studies were conducted to examine hatchling behavior and orientation. It was observed that if the slope of the nesting area was flat, hatchlings moved in all directions with respect to the nest chamber. However, there was a different emergence pattern for nests located on angled slopes. In the latter case, hatchlings tended to travel downhill, an orientation known as geotropism (although a few trudged up sloping dunes). These experiments in the field were also conducted under laboratory conditions on an artificial apparatus covered with sandpaper. It was found that the hatchlings used for laboratory testing displayed the same behaviors observed in the field. Hatchlings tended to walk down an incline (rather than up or to the side), although they hesitated for a longer period of time before moving on the artificial apparatus compared to field-based trials (Burger, 1977).

When we discover emerging hatchlings in the field, we often weigh and measure them as part of a population study. After they are handled and set back down to continue their journey, we often observe them for various periods of time until they have found cover. I have frequently noted that no matter what the slope may be, our released hatchlings scramble for the nearest hiding place, usually under *Spartina*, bayberry, or other low-growing plants. This often means an uphill trek. We may question whether the orientation toward vegetation and subsequent hiding is a response to being handled or if there is a strong adaptive component to this behavior. In a field orientation study conducted in the mid 1970s, vegetation was also a significant factor in determining the direction in which hatchlings traveled. After testing forty hatchlings in the field, Burger (1977) concluded that hatchlings gravitated toward the closest vegetation, no matter the type. These findings support other field observations in which hatchling tracks may radiate from the nest in a number of directions but disappear into the closest vegetation. Considering the fact that the hatchling's shell is still soft and that movement from the nest occurs during daylight hours, the fastest route to cover will be the safest one for the tiny hatchling.

Some seem to wander in looping patterns and are easily detoured by variations in beach contour presented by dunes, tire tracks, and even footprints. These detours and looping patterns may not be as random as they seem. Hatchlings may be responding to geotactic, olfactory, and other orientation signals that are poorly understood. Hatchlings may be attracted to the sand roadways because of the deposition of tidal wrack in this part of the marsh (Auger and Giovannone, 1979). The tendency to crawl within a tire track can be very unfortunate for some hatchlings. On Sandy Neck in Massachusetts, they have been found run over and killed in wheel ruts. A hatchling that was temporarily nicknamed "Lucky" was once found in a tire track within the Cape Cod National Seashore where it had been accidentally run over by a park ranger's truck. As its name implies, the hatchling survived the encounter with the vehicle, but not all hatchlings are this "lucky."

As hatchlings break out of their shells they are found in all orientations within the nest, including upside-down and sideways. As they make their way out, scrambling over their nestmates, they must be able to right themselves. Hatchlings sometimes flip upside-down as they climb dunes or tumble down slopes. Burger (1977) compared hatchlings that were more developed and at the verge of emergence to newly pipped ones. She found that the more developed hatchlings had an increased ability to right themselves. This ability negatively correlated with size of the yolk sac. The more developed the hatchling, the better it was able to flip itself over using its neck and limbs.

While it may be clear that newly emerging hatchlings employ a strategy of "prompt concealment," Coker (1920) marveled at the ability of the young terrapins to go into hiding. In two years of field observations, only fifteen terrapins with two growth rings (presumably two-year-olds) were found, while "not a single terrapin under two years of age rewarded the many careful searches of an experienced terrapin hunter, though the young must be more abundant than the older forms" (Coker, 1906). Coker also noted that on the few occasions on which young terrapins were found, they were within a small area of the marsh.

Additional studies suggest that hatchlings and juveniles may continue to seek cover in the marsh for up to three years (Coker, 1906). Pitler (1985) described a hiding behavior displayed by twelve terrapins in a New Jersey population. Over a three-year period, he actively looked for hatchling and juvenile terrapins with a carapace length of 2.5 to 7.5 centimeters (1 to 3 inches) in a tidal mud flat. The small terrapins were always discovered at low tide on well-drained substrate about 90 meters from the water. Hatchlings and juveniles were found primarily under matted *Spartina* and other low growing vegetation and also under rocks, debris such as wooden boards, and tidal wrack. Such hiding behavior protects these turtles from predator attack and offers shade for thermoregulation.

Lovich et al. (1991) studied the ecology and demography of diamondback terrapins in a South Carolina marsh. In order to study hatchling behavior, they first incubated terrapin eggs in the laboratory, then returned the resulting nine hatchlings (mean carapace length of 3.34 centimeters [1.3 inches]) to an area near their nest site and observed each hatchling for thirty minutes. All hatchlings avoided water. They all swam to land, even though observers were standing on shore, well within sight. Each hatchling burrowed under the wrack near the high tide line. This location may be ideal for hatchlings because it provides protective cover for these tiny turtles. In addition, the wrack provides shade and food. There is an abundance of small crabs, insects, and amphipods that share the wrack with hatchlings. Periodic tidal flooding also ensures moisture.

From the study just described, it was hypothesized that hatchling terrapins may routinely use microhabitat beneath tidal wrack. This led Lovich et al. (1991) to attempt to find them there. Despite intensive searches within the predicted microhabitat, the investigators found themselves looking for hatchlings in a salt-marsh haystack without success. A similar scenario occurred during a study of diamondback terrapins in a Delaware salt marsh. Despite a search, small terrapins were not found anywhere (Hurd et al., 1979). The investigators speculated that the missing age classes may be the result of catastrophic mortality.

On Cape Cod, we have also come up empty-handed when we sift through tidal wrack and search *Spartina* marsh for hatchlings. This does not mean they are not present. From various studies we now realize that inability to find young terrapins in the marsh may be due to their cryptic behavior. If we use special techniques for tracking hatchlings and juveniles, it is possible to recapture at least some of them. The inability to find representatives of the younger age classes is sometimes a matter of having too much area to cover. We occasionally come across a hatchling or yearling in the marsh when we are not purposefully seeking them.

Matt Draud has been more successful. When he and his students scour the marsh, meter by meter, they have found hatchlings and juveniles. One of the reasons that Draud has been more successful than others is that the swath of marsh bordering the Oyster Bay nesting area is very narrow, only a meter wide in some places. In contrast, the Wellfleet, Cape Cod, salt marshes extend

*Fig. 4.2. A radio has been attached to a headstarted terrapin so that its movements can be followed after its release.*

over many acres, with expanses of *Spartina patens* so thick that it is almost impossible to part the strands and reach bare substrate. Another reason that hatchlings and juveniles are difficult to find, even when a systematic effort is made to search the marsh, is that they frequently burrow under the mud, sometimes for long periods of time.

Draud traced the journey of Oyster Bay hatchlings and found that they move very little, perhaps 0 to 10 meters (0 to 11 yards), between recaptures. Hatchlings remain in the drier *Spartina patens* during the fall in which they emerge. As the weather changes and temperatures drop, the hatchlings move into a terrestrial habitat, slightly upland from the marsh. They hibernate in the dry uplands. Draud found that during the following spring they become more adventurous when they emerge from their first hibernation, displaying more movement than they did during their hatchling season. They move farther and farther as their first year progresses. Their movements follow a sequence from hibernaculum to wrack line to *Spartina patens* to *Spartina alterniflora*. During this period of movement, they are consuming insects, spiders, amphipods, and crabs. Draud found that the larger the yearling, the larger are the crab claws found in fecal samples. Movement is inhibited on windy days. Draud also found that the largest size disparity of yearlings occurred in late summer (Draud, and Bossert, 2002). Apparently, the young terrapins are not all growing at the same rate.

Wheaton College has partnered with investigators at Massachusetts Audubon's Wellfleet Bay Wildlife Sanctuary to study the movement and behavior of juvenile terrapins. We have used eight- to nine-month-old terrapins that have been headstarted in the laboratory. By feeding them all winter and keeping them active to prevent the first year of hibernation, the hatchlings achieve the size of two- to three-year-old turtles. They are comparable to juvenile terrapins; the males clearly display their secondary sex characteristics and some of the larger ones are approaching sexual maturity. After we attach small radio transmitters to their carapaces (fig. 4.2), they are released into their natal marsh at their original nest site. Most of these terrapins have a delayed response and will remain completely immobilized for several minutes. Like newly hatched terrapins, the first movements take them to the nearest vegetation, where they quickly tunnel under the grass or shrubs and remain in place for extended periods (plate 15). Many will burrow under the mud or substrate before moving more than a meter or two (3 to 6 feet) (fig.

*Fig. 4.3. Released headstarted terrapins often burrow under the mud.*

4.3). In more than forty headstart releases, we have observed one notable exception to the "stay put and hide" strategy. This little wanderer was a 154 gram (0.34 pound) male who traveled close to 200 meters (219 yards) upland and settled down in a wooded area of mostly pine trees on private property abutting the marsh. He remained burrowed in the same location for almost a month, under a thick layer of pine needle litter, chock full of worms, land snails, and insects. He acted more like a box turtle than a diamondback terrapin. He did not lose weight and appeared healthy throughout his stay in the pine uplands. Following a period of heavy rain, we lost his signal and did not see him again.

Our eight- to nine-month-old, lab-reared terrapins, equivalent in size to juveniles in the wild, remained at the site of their release for days, weeks, and sometimes months. By tracking fifteen of these terrapins with radio transmitters, we have been able to follow their activities in the marsh. Two young female terrapins remained in their natal marsh for the entire summer. The males were more adventurous; after remaining buried for several days to several weeks, they gradually ventured out into the larger creeks. When they reached the creeks, we could no longer detect their radio signals. Occasionally, one of them would reappear within the wetter areas of the marsh. With respect to overall movement, the general progression involved a period of time under *Spartina patens*, followed by periods buried in the mud in the *Spartina alterniflora* regions of the marsh. Our released terrapins each covered 1.5 to 3.0 hectares (approximately 4 to 7 acres) during July and August with a bit of overlap in the area of the marsh that they utilized. The longest observed trek for one of these juvenile-sized terrapins was 400 meters (437 yards) between recaptures. The greatest movements were observed during spring tides, when the entire marsh was inundated (Brennessel et al., 2004).

In areas with less developed *Spartina* salt marshes, such as the Patuxent River of Chesapeake Bay, hatchlings are rarely seen on land and juvenile turtles are usually not found in the marsh. To determine which part of the habitat might be utilized, Roosenburg et al. (1999) captured five juvenile males and seven juvenile females, as well as adult males and females. The terrapins were each fitted with a bobber on 4 meters (4.37 yards) of fishing line and tracked for three hours in midsummer from the site of release. The investigators discovered that juveniles of both sexes were found in the shallow, inshore water, while adult females used open water in the river and mouths of the creeks. Interestingly, adult males were found in the same shallow inshore waters as the juveniles of both sexes. The age and size of females correlated positively with distance from shore. It thus appears that a habitat shift from shallow to deeper water may relate more to terrapin size than to age or sex.

A similar trend is seen in other turtle species. In a study of habitat utilization by painted turtles (*Chrysemys picta*) and snapping turtles (*Chelydra serpentina*) in a Michigan marsh, Congdon et al. (1992) discovered a correlation between the size of the turtle and water depth. The correlation held for juvenile as well as sexually mature turtles. Young and small juveniles were found in shallow areas of the marsh, usually close to shore. Larger mature turtles tended to utilize shallow water in early spring but moved into deeper waters as the temperature increased. It was speculated that younger turtles prefer to remain in shallower habitats due to distribution of food resources, less developed swimming abilities, and predator avoidance. Deeper channels and tidal creeks have stronger currents than near-shore waters and would be more difficult for young terrapins to navigate on their own terms. Shallow water is also warmer, and a preference for warmer water could contribute to faster growth during this period when turtles are so vulnerable to predators. These reasons may very well explain the observations of shallow aquatic habitat utilization by young diamondback terrapins.

It thus appears that the movements of hatchlings and juvenile diamondback terrapins are initially confined to the well-drained marsh adjacent to their natal nesting areas. They seek cover under vegetation, wrack, debris, and/or mud. As they become more mature, they venture into wetter sections of the marsh, but still burrow under mud for long periods. The most mature of the juvenile group use marsh creeks and near-shore shallows but may return inland periodically.

## Food Preferences

Observations of diamondback terrapins in the early 1900s led herpetologists to believe that hatchlings in the wild do not eat until the spring after their emergence. This appears to be the case, especially in northern geographical areas where hatchlings found in spring are the same average sizes as those discovered the previous fall (Coker, 1906). Some of these hatchling may have overwintered in their nests.

In a healthy marsh, there is an abundance of food. What do hatchlings and juveniles eat? A laboratory study of the feeding habits of 875 immature diamondback terrapins may shed some insight into the food preferences of young terrapins. A cafeteria-style experiment was set up in which the terrapins, weighing between 4.0 and 7.85 grams (0.14 to 2.8 ounces), were offered crabs, oysters, clams, marsh snails, mussels, fresh fish, canned fish such as tuna and salmon, and meat such as raw steak and ground beef. Initially all food items were accepted, but after three weeks the canned fish was ignored. Liver was not a preferred food but it was tried occasionally. Shellfish and snails were the most preferred items, followed by fresh fish, crabs, and beef. Some of the hatchlings, 14.06 percent, refused to eat (Allen and Littleford, 1955). An interesting outcome of this laboratory study was the finding that young terrapins can become very selective in their food choices after a few weeks. Once preferences were established, it proved very difficult to produce a change in feeding patterns.

In our pilot headstarting program at Wheaton College, we observe an initial lag period after the yolk sac is resorbed in which hatchlings will not eat, no matter what types of food we offer. However, within a few weeks, most hatchlings will be feeding on commercial hatchling food, supplemented with brine shrimp, fish, scallops, clams, or mussels. In a group of thirty to forty hatchlings, there are always one or two that will not feed during the first few weeks to several months, or feed so little that they don't gain weight. Usually the hatchlings that are slower to feed begin to actively feed and grow much later, sometimes after their tankmates have tripled or quadrupled in size. These smaller hatchlings will experience the same type of growth spurt as their tankmates when they actively feed, but do not reach the same size by the time they are ready for release in spring.

In the field, we can collect fecal samples from hatchlings and juveniles to examine their dietary preferences. Our tracking studies of released headstarted terrapins in Wellfleet (Brennessel et al., 2004) and Matt Draud's tracking studies of hatchlings and juveniles in Oyster Bay (Draud and Bossert, 2002: Draud, Zimnovoda, King, and Bossert, 2004) display similar trends. The young turtles are sampling various food items in the marsh substrate, including small crabs, salt-marsh snails (*Melampus*), various insects, and marsh invertebrates. Similar to more mature terrapins, they appear to be opportunistic carnivores. Researchers have not been able to identify any specific food preference in the young population, although the smaller hatchlings rely heavily on the salt-marsh snail, *Melampus*, a gastropod found in the intertidal region.

## Hatchling Hibernation

Diamondback terrapins hibernate during winter throughout much of their range, even as far south as Florida (Butler et al., 2004). In contrast to adults who hibernate in muddy sediments of creeks and coves (chap. 1), hatchlings hibernate on land. The environment within the hibernaculum will determine the outcome of this winter dormancy period. As mentioned previously, some terrapins never leave their nest and spend their first year in hibernation, a phenomenon known as overwintering or delayed emergence. This strategy is also seen in other turtle species. In cases in which hatchlings overwinter, the placement of the nest by the mother turtle will impact not only sex of hatchlings and their development period, but also their winter survivorship. Overwintering is an adaptation that allows for conservation of energy. The hatchlings stay put over the winter and emerge with warming temperatures and an abundance of food. Overwintering on land is an adaptation that can also protect hatchlings from fall predators and eliminate the types of osmotic and hypoxic stress that can be encountered during aquatic hibernation. There is a downside to overwintering: The soil temperature can sometimes dip below freezing for significant periods of time. It is not an easy way to avoid stress because it creates a new form of stress. It can be a perilous period in which the tiniest hatchlings may be subjected to extreme cold temperatures and to flooding of nests due to fall hurricanes and winter storm surges. The overwintering hatchlings must go without food and sometimes without water if the ground around the nest is frozen. Perhaps this explains the mixed emergence strategy characteristic of terrestrially hibernating hatchlings: Some hatchlings from a clutch may emerge in the fall, while others remain until the following the spring. Each season presents its own hurdles for survival.

Although we do not have accurate information about how many hatchlings survive their first hibernation, data from early twentieth-century terrapin farms in North Carolina indicate a 13 percent death rate in hibernating stock, compared to a 6.5 percent death rate in hatchlings that were kept active all winter in a hothouse (Coker, 1920). The first hibernation may be the most perilous.

The ability of hatchling turtles to survive hibernation presents a physiological puzzle. Water is not available when the ground is frozen, and the small size of hatchlings makes them particularly prone to desiccation. However, lack of water may be the least of their winter stressors. With the possibility of fluctuating winter temperatures that may dip below freezing for extended periods of time in northern areas, one wonders why hatchlings don't freeze to death. The parts of the hatchling that are exposed to the outside environment, structures such as eyes, nares, integument (skin), cloaca, and unhealed umbilical scar, are particularly prone to freezing. Naturally produced antifreeze compounds known as cryoprotectants have been found to protect other organisms from converting their body water to ice at extreme cold temperatures. Emydid turtles do not appear to produce any type of antifreeze molecules. They can potentially respond to cold temperature stress in two ways. The first method is by supercooling and remaining unfrozen at low temperatures. Organisms that use this strategy will tend not to freeze unless there is an ice nucleating agent, such as sand or bacteria, present in their systems. Other organisms survive freezing by becoming freeze tolerant. They convert some of their body water to ice but somehow survive periods of being frozen like a popsicle. Land-hibernating terrapin hatchlings use the latter strategy. Baker and his colleagues have examined the freeze tolerance of diamondback terrapin hatchlings for short periods of time in efforts to understand their cold hardiness. They subjected hatchlings to below-freezing temperatures for various periods of time and assessed their ability to survive. They discovered that hatchlings tolerate short periods of freezing under conditions that mimic those found in nature. For example, when hatchlings were cooled to -2.5°C (27.5°F) for fifty-three to ninety-seven hours, twelve out of thirteen survived. When held at the same temperature for a week, seven out of seven survived. But Baker's group found clear limits to the ability of the hatchlings to recover from a frozen state. When the freezing period was extended to twelve days, none of the hatchlings survived. Furthermore, when hatchlings were cooled to −3.5°C (25.7°F), 75 percent of body water was converted to ice, and after eighty hours, no hatchlings survived. In this study, no differences were found in the cold tolerance between New Jersey and New York hatchlings (Baker et al., 2004). The limitations in cold tolerance found in these laboratory freezing experiments help to explain why terrapins are not found north of Cape Cod where more prolonged periods of ground freezing are typical.

## Hatchling Cycles/Circadian Rhythms

On the north shore of Long Island, a diamondback terrapin population was studied in marshes and creeks in the town of Cutchogue (Muehlbauer, 1987). Freshly laid eggs were harvested and incubated in the laboratory so that hatchling activity could be monitored under controlled conditions. By record-

*Plate 1.* *Sexual dimorphism: the larger female* (left) *has a shorter, thinner tail than the smaller male* (right).

*Plate 2.* *Diamondback terrapins in a peeler tank at the Horsehead Wetlands Center. Variation in carapace design and color can be discerned.*

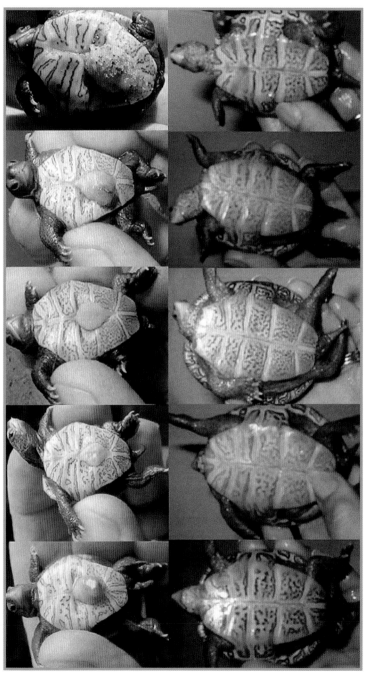

***Plate 3.*** *Unique plastron markings can be used to distinguish hatch-lings and juveniles. The markings lose their definition as terrapins age. Hatchling photos by Don Lewis.*

*Plate 4.*
*Wellfleet Harbor*
*in winter.*
*Harbor, creeks*
*and Herring*
*River Estuary*
*may remain*
*frozen for*
*several months*
*each winter.*

*Plate 5.* *Wellfleet, Massachusetts, salt marsh vista.*

***Plate 6.*** *Gravel substrate in Oyster Bay, New York, nesting area.*

***Plate* 7.** *A female has her eggs eaten by a raccoon as she deposits them in her nest.*

*Plate 8.* Habitat of the mangrove terrapin. Arched prop roots of red mangrove and upright pneumatophores of black mangroves within the marl substrate. Female terrapins were found buried within pneumatophores under the large black mangrove.

*Plate 9.* Oval eggs have a pink tinge when they have been freshly laid. Photo by Don Lewis.

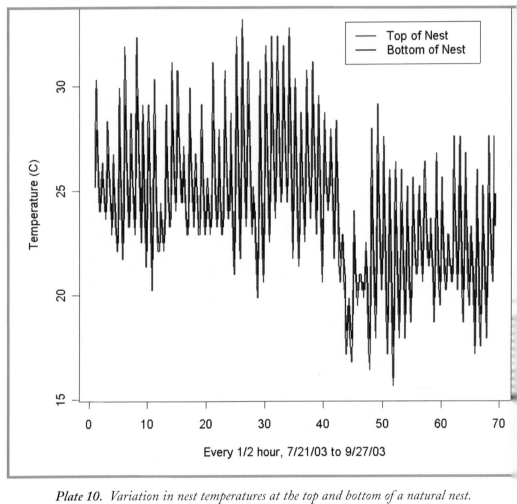

*Plate 10.* Variation in nest temperatures at the top and bottom of a natural nest.

***Plate 11.*** *Egg tooth is visible on the beak of this newly hatched terrapin. Photo by Don Lewis.*

*Plate 12.* *Hatchlings emerging from eggs. Photo by Don Lewis.*

*Plate 13.* *"First Breath" sequence photographed by Don Lewis.*

*Plate 14.* *Emerging hatchling leaves tracks on sand. Photo by Don Lewis.*

*Plate 15. Released headstarted terrapins quickly seek cover.*

***Plate 16.*** Turtles Dancing. *Drawing in color by Theresa Drift. Mikinaakwug Miimi"idiwug* © *1999–2006 Theresa Drift. (Used with permission.)*

ing electronic activity to monitor locomotor activity in individual hatchlings, it was shown that hatchling activity correlated well with the twelve-hour pre-set light cycle. Hatchling activity was lowest during the twelve-hour dark phase. Furthermore, in three of the five hatchlings studied, there were smaller activity peaks and lulls that alternated, with a mean period of 5.17 hours. This latter observation suggests a possible innate activity cycle that is linked to tides. In a study conducted at Wheaton College in which the daily activities of four hatchlings were monitored using web cameras attached to a computer that was programed to analyze locomotor activity, we found that hatchlings can actually anticipate the photoperiod induced by laboratory lighting. The hatchlings begin to display locomotor activity about thirty minutes before the onset of light (Hunt et al., 2002). During the light period, there is also the suggestion of a slight underlying activity fluctuation at approximately six-hour intervals. If these six-hour fluctuations turn out to be significant, they would agree with Muehlbauer's observations. We can then can ask: Why should terrapin hatchlings display a diurnal activity cycle with tidal variation?

If hatchlings are active primarily to find food, tide-related activity would be a beneficial adaptation. High tide would provide more protection from overhead avian predators and some diurnal terrestrial predators compared to low tide, when more of the marsh is exposed. These laboratory findings also agree with our field observations of released headstarted terrapins that we find burrowed under the *Spartina* and marsh mud during low tide but display larger movements during high tides.

## *Headstarting*

It seems clear that tiny diamondback terrapin hatchlings, about the size of a quarter and weighing about 5 grams (0.18 ounces), are vulnerable to predators and the elements. Would hatchlings have increased chances of surviving if they could achieve a larger, predator-proof size in a shorter period of time? This is the rationale for some of the programs that are designed to provide individual turtles with a head start in life.

For diamondback terrapins, the incentive to headstart was initially eco-nomic. Programs were initiated in response to the commercial demand for turtle soup. As a result of a suggestion by H. B. Aller in 1911, while he was superintendent of the U.S. Fisheries Biological Laboratory at Beaufort, North Carolina, the first year of terrapin hibernation was eliminated in order to increase terrapin survival and to jump-start growth. The theory behind this experimental strategy is that young turtles will have a better chance of survival if they can spend their most vulnerable developmental period, when predation rates are expected to be very high, in a protected environment. Furthermore, if growth of terrapins could be accelerated during headstarting, especially the wasteful period of the first winter when terrapins do not feed and thus do not grow, a new generation of turtles could be launched in five or six years rather than six or seven. This acceleration in growth could potentially supply the commercial demand for terrapins at a more rapid rate. Thus, the Fisheries Bureau housed hatchlings in a modified greenhouse maintained at 36°C (80°F ) while keeping them active and feeding them throughout the winter. The headstarted terrapins reached the size of two- to three-year-olds within a matter of months.

Headstarting is not unique to diamondback terrapins. The strategy has also been used as a method to restore and preserve threatened and endangered marine turtles such as the Kemp's Ridley, Hawksbill, and Green sea turtles. Headstarting has also been employed in attempts to improve chances of survival for freshwater turtles such as the Western pond turtle (*Clemmys marmorata*) listed as endangered in Washington State, most likely due to exploitation for food, and the Plymouth red-bellied turtle (*Pseudemys rubriventris bangsi*), endangered as a result of development of shoreline and upland habitats surrounding ponds in the restricted region of Massachusetts where the turtle is found.

Turtles that will be headstarted can be obtained in a number of ways. In some cases, eggs are removed from natural nests and incubated in an artificial environment to produce hatchlings. In other cases, hatchlings may be obtained from natural or protected wild nests at the time of their emergence. Under circumstances in which turtles are "farmed," a captive breeding population is maintained and offspring from these adults are raised.

Sea turtle headstarting programs were first initiated in the 1950s due to conservation concerns, and the practice continued in the 1960s for added economic purposes. A tremendous amount of controversy surrounds these efforts. In the case of sea turtles, saltwater containment tanks or enclosures are required. Specialized equipment is used to keep the water clean and to regulate temperature. The cost to maintain a sea turtle farm is often prohibitive, and these operations have not been found to be cost-effective in producing turtles for the commercial food market. It is not just a matter of expensive equipment; it takes many years before a turtle achieves a marketable size.

Other problems plague the turtle farmers. These plagues are caused by parasites and microbial pathogens that may be present a low levels in natural populations. In crowded holding tanks, a disease may spread quickly through the captive animals and cause a high rate of mortality. There are also concerns about releasing potentially infected animals back into the wild population, where devastation of natural populations may result.

It had initially been proposed that the demand for commercial turtle products, including eggs, oil, shell, and meat, could be met by use of farm-raised turtles. This would potentially decrease the hunting of wild turtles and poaching of eggs from natural nests. However, there is some fear that turtle farming may reopen the international trade in sea turtle products and that turtle farms may not be able to keep up with the demand. This may lead to an increase in the illegal capture and sale of turtles and their products.

In the case of using headstarting as a conservation strategy, various issues must be kept in mind. Questions have been raised about the ability of head-started turtles to adjust to life in the wild. Some studies suggest that head-started turtles eventually join the breeding population, but other research suggests that headstarted animals do not behave in a normal manner. Certain survival behaviors may be learned during the early period of a sea turtle's life. In their juvenile years, sea turtles develop complex migratory behaviors that may be altered when a turtle is raised under artificial conditions. Perhaps the biggest concern with headstarting sea turtles and releasing them to attempt to increase or stabilize natural populations is that there is no indication that the strategy may be successful.

Whether the issue is conservation of threatened or endangered populations or economic viability in terms of producing turtle products, the jury is still out on sea turtle headstarting. The only proven arena in which headstarting has been a success is tourism. The Cayman Turtle Farm is a major attraction, and tourist dollars help pay the bills.

In the case of freshwater turtles, there may be fewer problems with husbandry. Expensive saltwater systems are not required for maintenance of the hatchlings, and there is less concern about the development of the ability of headstarted turtles to migrate because freshwater turtles tend to remain in localized populations. The State of Washington Department of Fisheries and Wildlife initiated a recovery program for Western pond turtles (*Clemmys marmorata*) with a headstarting component. The agency reports that the Western pond turtle now numbers 250 to 350 of which half originated as headstarted turtles from the Woodland Park Zoo (http://wdfw.wa.gov/wlm/diversty/soc /recovery/pondturt/wptxsum.htm). When the data clearly show that these introduced turtles become part of the breeding population, there will be further evidence for the success of the program.

With respect to headstarting, diamondback terrapins are more similar to fresh water turtles than their marine counterparts. Pilot terrapin headstarting programs have provided guidelines for successful husbandry. These programs have been initiated for a number of reasons. The largest program I have seen involves a collaboration between the Wetlands Institute in Stone Harbor, New Jersey, and Richard Stockton College of New Jersey. Roger Wood, director of the Wetlands Institute, and a professor at Stockton, has enlisted the assistance of interns and faculty for this project. Hundreds of tiny hatchlings fill plastic bins in the animal facility at the college. They originate from eggs dissected from the carcasses of females that are killed on the causeways leading to the Jersey Shore. From the carnage, Roz Herlands, a developmental biologist and professor at Stockton, rescues the eggs and incubates them at 30°C (86°F), a temperature that will produce mostly females. The rationale in producing females is to replace the road-killed mother turtles that carried these eggs, many of which may still be intact and viable. Herlands has optimized the procedures for these emergency oophorectomy operations (surgical removal of eggs), the transport of the delicate eggs to the college, and the conditions of incubation so that her success rate for producing hatchlings is around 70 percent. In the first year of the program, much to the dismay of the egg rescue team, hatchlings were scooped up by gulls when they were released in the wild. Wood recalled that from the bird's perspective, the hatchlings were bite-sized snacks. After all the effort to remove eggs and incubate them, it was very disappointing to see all the hatchlings gobbled up in short order. If the goal was to "replace" the road-killed females, something clearly had to be done. So Wood and Herlands began to headstart the hatchlings and to plan their release for the following spring. Their program has been in place since 1997, and some of their many baby turtles were cared for at the Philidelphia Zoo. When larger turtles are released, gulls pose less of a problem. The program appears to be working; the collaborators are beginning to find some of their headstarted animals mingled with the Cape May juvenile and adult population.

The Turtle Tots program was initiated by the Maryland Fisheries Service Department of Natural Resources (DNR) as part of a larger outreach project called Terrapin Station, designed to promote environmental awareness and get the public more involved in issues of resource management and habitat protection. The Maryland DNR selected the diamondback terrapin, the mascot of the University of Maryland and the state reptile, as a hook to get the public interested in the resources of the Chesapeake Bay watershed and keep the public informed about conservation issues. There is more about the project in chapter 6. The program had a headstarting component using rescued eggs from gardens and shorelines around Chesapeake Bay that were deemed inadequate for nesting success. The eggs were incubated and the resulting hatchlings were shipped out to schools where they were tended to by eager children and their teachers throughout the winter. After eight or nine months, the much larger hatchlings were released after being fitted with wire tags, so that if they were caught the DNR would know they were graduates of Turtle Tots. The program became very popular; at one point, eighteen schools were raising baby diamondback terrapins and there was a growing demand for more hatchlings. Maryland Governor Parris N. Glendenning established a Terrapin Day, on which school-raised terrapins were released into the wild with great fanfare. Although the program was tremendously successful with students and teachers alike, there were some lingering concerns. Would interactions with humans somehow deprogram the terrapins from exhibiting normal behaviors in the wild? Would the terrapins pick up pathogens from their human caretakers and pass them along to the wild population? Would the school kids become infected with *Salmonella* from the turtles? With these troubling questions, decisions were made to scrap the headstart program and put resources into other initiatives to raise environmental awareness.

On Cape Cod, Peter Auger began to headstart terrapins from tire-track nests. These nests were doomed to be crushed by off-road vehicles such as trucks and sport utility vehicles (SUVs) as drivers made their way out to Sandy Neck for fishing and recreation or to reach a small cottage colony at the tip of the barrier beach. Auger began collecting the eggs from these nests and incubating them at Barnstable High School. Rather then seeing the hatchlings become snacks for predators, he began to keep them over the winter and release them, about forty times larger than their hatchling size, the following spring. He also realized the value of these small turtles as teaching tools and placed some of them in classrooms where students acted as caretakers, observed their behaviors, and charted their growth. Concerned about their adjustment to their natal marsh after being fed tasty treats and vitamin-supplemented pellets, he created small pens within the marsh where they were kept for a period before their release. By observing them in the enclosures, Auger was somewhat assured that the headstarted terrapins behaved like their normal counterparts. Despite the artificial environment for the terrapin, the whole experience seems a better fate than being squashed by a tire.

My experience with headstarting also began as a rescue of sorts. A fall storm on Cape Cod eroded a sand dune and exposed a clutch of terrapin eggs. In case they were viable, Don Lewis rescued the eggs and put them in a bucket of moist sand and brought them into his garage. One day, Lewis looked into the bucket and he was mildly surprised to find five healthy hatchlings. But by the time the hatchlings emerged in the warm garage, the outside temperatures had plummeted and the late fall winds were whipping. Cape Cod terrapins had already disappeared for the season and were most likely in hibernation mode. A release into the marsh would have been a death sentence for the hatchlings. Our solution was an eight-month vacation in a nice warm aquarium. Similar to other groups that have headstarted terrapins, we were also concerned about the fate of these turtles and their adjustment to the marsh. This was our incentive to track their movements after they were released. To assist with the transition to the wild, the hatchling diet of commercial pellets was supplemented with fresh marsh food such as mussels and periwinkle snails during the month prior to their freedom. After the hatchlings were released with radio transmitters attached to their carapaces, we periodically recaptured them, recorded their weights, and checked fecal samples to assure ourselves that they had made the switch from a diet of hatchling pellets to various marsh offerings.

The headstarting process has revealed additional information about the environmental conditions that can potentially impact hatchlings in wild populations. It is possible to raise terrapin hatchlings in fresh water, but most headstarting protocols call for the addition of salt into aquaria. The salt inhibits growth of unwanted bacteria and also helps to prepare hatchlings for their natural environment. Hatchlings may experience full-strength seawater in some of their natural marsh habitats, but in captivity they appear to be most healthy and have better survival when they are started off under very low salt concentrations, 3 to 5 ppt, and later reared at salt concentrations from 5 to 10 ppt.

It is no surprise that temperature is important to hatchling growth and that hatchlings raised in aquaria or tanks maintained at approximately 26 to 28°C (80 to 83°F) grow more rapidly than those maintained at lower temperatures. They must be provided with lighting that mimics daylight, with proper ultraviolet wavelengths for shell hardening and development. Terrapin hatchlings resemble pond turtles when they pile up on top of one another to bask, one of their favorite activities.

One of the husbandry problems identified from headstarting conducted at Beaufort in the early 1900s was a condition known as softshell (also described in chapter 1 in relationship to vitamin D deficiency). This disease followed from a failure of the hatchlings to eat, with a resulting failure to grow. Without active growth, the hatchling does not incorporate the minerals needed for the shell to harden. Softshell animals have a high mortality, although some start to feed at a later time and can recover from the condition (Hildebrand, 1929).

A surprising terrapin behavior encountered in most headstarting situations is aggression. Hatchlings that are housed together in the same tanks may experience a change in behavior from peaceful coexistence to a situation in which one or more begin to attack others by biting and chewing, mostly on tails or rear limbs. The resulting injuries usually heal but occasionally become infected and may lead to permanent deformities. Aggressive hatchlings should be separated from others. We sometimes use tank dividers to keep the peace. The cues that promote aggressive behavior are not well understood, but one situation that frequently triggers aggression is overcrowding. If a hatchling is injured, attacks will continue. The aggressor is usually joined by other tankmates, perhaps because of their attraction to the open wound.

Despite the anecdotal success of headstarting efforts for diamondback terrapins, the strategy must still be considered as experimental. It is too soon to determine whether headtstarted diamondback terrapins will mate and whether females will nest. We also don't know if wild or headstarted terrapins return to their natal beaches to nest, a phenomenon first described in sea turtles by Archie Carr and known as natal homing. Using mechanisms and cues that are still a mystery, sea turtles "imprint," that is, learn a behavior or response that is usually irreversible, on their natal beaches. The imprinting is believed to occur in a short but sensitive period and may occur when turtles are developing in the nest or as they make their way to the ocean as hatchlings. We don't know if wild diamondback terrapins return to their natal beaches, so it is too soon to assess the ability of headstarted/released terrapins to return to home shores to nest.

Headstarting of diamondback terrapins is thus in its investigational stages, and more work must be done to determine whether this strategy is reasonable and defensible. Decisions may need to be made on a case-by-case basis. Headstarting may have the potential to contribute to conservation/ repatriation efforts if it can be shown that headstarted terrapins do not carry or transmit pathogens, do adjust to life in the wild, and eventually reproduce. At the very least, headstarted terrapins that would otherwise not have survived have a tremendous public relations and educational value. Cute little turtles provide a visual mechanism to teach about terrapins and draw people into the concept of conservation. An added benefit is that headstarting will continue to inform us about behaviors and adaptations of young terrapins.

Although we still cannot fully characterize their lost years, we can speculate that hatchlings disperse from their nest sites and most likely live a solitary existence in the wild. They spend much of their early years in a cryptic mode and remain in the marsh and smaller tidal creeks for several years before venturing into deeper water. Their encounters with other young terrapins may occur by chance, but these meetings are probably infrequent. Social behaviors and terrapin–terrapin interactions are exhibited after sexual maturity and occur during mating. Research efforts should be designed to focus on early terrapin life, a period during which terrapins appear the most vulnerable.

# Chapter 5

## A Clear and Present Danger for The Most Celebrated of American Reptiles

### HOW THE PARTRIDGE GOT HIS WHISTLE

In the old days, the Terrapin had a fine whistle, but the Partridge had none.

The Terrapin was constantly going about whistling and showing his whistle to the other animals until the Partridge became jealous, so one day when they met the Partridge asked to try the whistle.

The Terrapin was afraid to risk it at first, suspecting a trick, but the Partridge said, "I'll give it back right away, and if you are afraid you can stay with me while I practice."

So the Terrapin let him have the whistle and the Partridge walked around blowing on it in fine fashion.

"How does it sound with me?" asked the Partridge.

"Oh, you do very well," said the Terrapin, walking alongside.

"Now how do you like it?" said the Partridge, running ahead and whistling a little faster.

"That's fine," answered the Terrapin, hurrying to keep up, "but don't run so fast."

"And now, how do you like this?" called the Partridge, and with that he spread his wings, gave one long whistle, and flew to the top of a tree, leaving the poor Terrapin to look after him from the ground.

The Terrapin never recovered his whistle, and from that, and the loss of his hair, which the turkey stole from him, he grew ashamed to be seen, and ever since he shuts himself up in his box whenever anyone comes near him.                                    —Cherokee legend

TURTLES HAVE AN important status in many cultures (plate 16), and they are major characters in folk legends around the world. Many species have been the subject of stories, fables, art, medicine, and magic. It is often unclear which type of turtle is being described. For example, "terrapin" was a generic term for several different species of turtle and does not necessarily refer to diamondback terrapins. Nonetheless, these "terrapin" stories are interesting as representative examples of human respect and reverence for turtles. The Cherokee legend describing how the partridge got his whistle is a bit unusual because the terrapin in this story is not as wise as turtles are usually portrayed. In contrast, an African folk legend recounts how the slow ambling gait of the tortoise caused overconfidence in the hare who challenged the tortoise to a race. The racing theme also emerges in other legends describing contests with fleet-footed animals that are won by the slower but wiser turtle.

When Joel Chandler Harris collected Southern plantation tales and retold them in his famous Uncle Remus stories, Brer Tarrypin was a prominent protagonist. Harris describes the terrapin in slave dialect. "Brer Tarrypin kare his house wid 'im. Rain er shine, hot er cole, strike up wid ole Brer Tarrypin w'en you will en wilst you may, en whar you fine 'im, dar you'll fine his shanty" (Harris, 1930). Brer Tarrypin usually managed to outwit Brer Rabbit, as he did in the story "Mr. Rabbit Finds His Match At Last." In this legend, Brer Rabbit bragged that he could easily catch Brer Tarrypin, so a fifty-dollar bet was struck and a five-mile race was planned. Clever Brer Tarrypin arranged to race in the woods instead of on the road with Brer Rabbit. Everyone thought he was being foolish, but Brer Tarrypin had a wife and three children and they were all "de ve'y spit en image er de ole man" (Harris, 1930). No one could tell them apart. On the day of the race, the Tarrypin family spread out through the woods and took up their positions at the mile markers. At each milepost, Brer Rabbit came across a turtle whom he mistook for Brer Tarrypin. He ran faster and faster to the finish line, only to see that Brer Tarrypin had already arrived and was collecting the fifty-dollar wager.

A very similar account is the subject of a Native American story from the Seneca tribe describing Turtle's race with Bear. In the Seneca legend, Bear and Turtle are racing near a frozen pond. Bear runs along the banks while Turtle swims under the ice. Instead of mileposts, holes in the ice mark the progress of the race. Turtle periodically pops his head out from the holes, showing Bear that he has the lead. When Bear comes to the finish line, Turtle is already waiting for him. Bear goes home tired and sleeps until spring.

*Fig. 5.1.* Feeding and Catching Terrapin on a Maryland Farm. *1888.*
*Benjamin West Clinedinst. Courtesy of Sterling and Francine Clark Art Institute,*
*Williamstown, Massachusetts.*

After Bear leaves, Turtle taps the ice and a dozen turtle heads emerge. Turtle thanks his relatives for helping him outwit Bear.

The most important turtle of all to Native American tribes was Mother Turtle, who carries the earth on her back. On the Objiwe tribal flag, a turtle symbol is placed prominently in the middle, where it represents Mother Turtle who cares for all her children. In their creation legend, Mother Turtle emerged from the water with earth on her back, providing a place for all her children to live between sky and water.

In Iroquois creation stories, the beginning of human civilization was traced to the time when a pregnant Skywoman fell to an island formed by the shell of a giant turtle. Turtle ordered all the other animals to bring up mud from the bottom of the water and place it on his back to form the land. When Skywoman gave birth, her progeny spread out over the land formed on Turtle's back.

In another Native North American creation story, two turtles were involved. The Earth was on the back of Great Turtle but the sky was dark. Little Turtle was sent to the heavens to get some light. She proceeded to collect lightning bolts that she formed into two balls. The large ball became the sun, and the small ball became the moon.

In his summary of diamondback terrapin cultivation efforts, Coker sums up a general feeling that people have about turtles. "It is little to wonder at then, that mythologists and fabulists have thought to divine in the tortoise, beneath a taciturn demeanor, inexpressive dome and inscrutable countenance, a shrewd and super-animal intelligence, or even a sense of cosmic responsibility" (Coker, 1920, p. 172).

The finding of diamondback terrapin shell fragments in shaman medicine kits indicates that in some Native American cultures the terrapin may have had a special status. Shell fragments may have been sacred objects or may have held magical powers. However, the relationship between diamondback terrapins and humans has not always been one in which respect and reverence for turtles has led to the protection of this species. If we examine human interactions with the diamondback terrapin we see that we are this turtle's main predator. In examining archeological sites along the Atlantic coast, it is evident that diamondback terrapins were plentiful when native peoples settled these areas. It should be no surprise that this tasty turtle became part of the diet. Examination of shell remains found in middens (great mounds of shells and "kitchen" trash, covered by dirt) of the Piscatways of southern Maryland indicates that in addition to clams, oysters, and other shellfish, natives hunted and harvested turtles, including the diamondback terrapin. Middens found along the South Carolina and Georgia coasts also contained terrapin remains.

## The Diamondback Terrapin's Rise to Culinary Fame

The terrapin gets its name from Native American sources. In the 1600s, it was called "torope" by Virginia Algonqiuans, "turepe" by the Abenakis, and "turpen" by the Delawares. Roughly translated, the name means edible or good tasting turtle. When European colonists arrived and founded settlements in Maryland and Virginia, the terrapin was found in abundance. In the late 1500s, Englishmen, led by Sir Walter Raleigh, who were exploring the bounty of the New World in a region that is currently part of North Carolina, described some of the bounty in this region of the country. Turtles and terrapins were specifically mentioned as a source of good meat, and their eggs were also cataloged as a resource. Early settlers most likely learned to cook terrapin from the natives, who prepared them as they would other shellfish: buried live in a bed of hot coals. When terrapin were abundant, large quantities of them could be easily netted as they basked on the surface of warm shoals. Some stories recount that they were so plentiful they were fed to pigs.

Terrapins, an inexpensive source of nourishment, were fed to servants and slaves by landowners in the Chesapeake Bay region. This practice was apparently the cause of a slave rebellion in the 1700s. The slaves were fed so much terrapin that they rose up to demand more diversity in their diet. In various writings, there are references to a 1797 Maryland statute that required landowners to limit the number of days they could feed terrapin to indentured servants, but Maryland officials have not been able to document such a law.

Many anecdotes and pictures attest to the former abundance of diamondback terrapins along the mid-Atlantic coast. There are reports of terrapins being so numerous in North Carolina that they were an annoyance to fishermen. Terrapins would fill up fishing nets and make them so heavy that the men could not draw them in, resulting in the loss of fish (Coker, 1906).

In colonial times, a wagon load of terrapins could be purchased for one dollar. By the early 1800s, food preparations made with the diamondback terrapin assumed a higher culinary status, and terrapin soups and stews became gourmet specialties. The fisheries status of the terrapin changed from annoyance or by-catch to a valuable commodity. Subsequently, the price of terrapin increased in a dramatic fashion. From $4 a dozen in Galveston Bay or $6 a dozen in Maryland in the early 1800s, some terrapin fisheries were able to demand over $100 a dozen for their largest terrapins by the 1850s. The development of a full-scale terrapin industry can be traced to the Chesapeake Bay region of Maryland. Watermen harvested terrapins and wholesalers collected and shipped them to markets and restaurants. In general, terrapins were sold by size with the largest specimens demanding the highest price. By the early 1900s, commercial grades and market prices for terrapins were as follows:

Counts: over 8 inches plastron length: $96–125/dozen
7–8 inches plastron length: $60–70/dozen
6–7 inches plastron length: $35–40/dozen
Half Counts: between 5–6 inches plastron length: $20–25/dozen
Bulls (males): $10–12/dozen (Coker, 1920)

The method of fishing for terrapins was somewhat dependent on the season. In summer, when these turtles forage in creeks and marshes, men and boys would wade in the shallows and catch them bare-handed or with hand-held nets (fig.5.1). Females would be captured when they left the waters to lay their eggs. Dogs were trained to find them on land and to track them as they returned to the water. Experienced "tarpinners," as terrapin fishermen were sometimes called, developed techniques to capture terrapins in cooler weather, after they burrowed under the mud. Long poles were used to strike or tap the muddy bottom layers of shallow creeks and marshes. The hearing of a "thud" accompanied by resistance to the force exerted on the pole usually signaled the presence of a buried terrapin. A more high-tech method for terrapin capture utilized baited traps that were positioned to allow the terrapins to come up to the surface to breathe, thus preventing their death by drowning. Terrapin drags were also fashioned. These devices allowed watermen to harvest terrapins from their boats by raking the mud and scooping up terrapins from the bottom layers. As the commercial and seasonal value of terrapins increased, Chesapeake watermen would catch terrapins during the summer months and impound them in dark, cool storage areas so that they could be sold for higher prices during the winter.

Terrapin soup appeared on the menus at upscale eating establishments, hotels, and private clubs. Terrapin became the first course at meals served during gala events. In 1862, on behalf of a supporter of President Abraham Lincoln, Joseph E. Segar shipped "a package of 2 dozen terrapin—a favorite luxury of my section of the country" to the White House for the President's enjoyment.

In 1877, *Schribner's Monthly* published an article on "Canvas-Back and Ter-

*Fig. 5.2.* Canvas-back and Terrapin; Dividing the Spoils. *Courtesy of Cornell University Library, Making of America Digital Collection. Cover of* Scribners Monthly; An Illustrated Magazine for the People, *Vol. XV, No. 1, November, 1877.*

rapin." The article extolled the culinary virtues of the two species, provided hunting and fishing tips, and described the economic importance of the species around Chesapeake Bay (fig. 5.2).

The gustatory fame of the "bird," as the terrapin was sometimes called, extended beyond the Chesapeake. The *Princeton Press* of January 29, 1887, reported the opening of a new restaurant and the celebration of the birthday of its proprietor: "Moses Schenck celebrated his birthday, the 54th, and the opening of his new restaurant on Hulfish Street, on Wednesday evening, by a Dinner party. The bill of fair [sic] was elaborate." The meal began with turtle soup and terrapin and was followed by such delicacies as roast turkey, escalloped oysters, beef à la mode and cold tongue. "The party discussed the Menu for two hours, decided that Mr. And (sic) Mrs. Schenck knew how to keep a restaurant, and retired wishing them the best of luck in the management of their business in its new location."

By the mid 1800s, the terrapin had been harvested for commercial use from Chesapeake Bay, from Galveston Bay, and from fisheries in North Carolina. Smaller scale terrapin harvesting also occurred in Delaware, New Jersey, New York, Connecticut, and the Cape Cod area of Massachusetts. Terrapins from Pleasant Bay on Cape Cod were packed in barrels and shipped to markets in Boston and New York. An enterprising fisherman from Barnstable, Massachusetts, caught and sold terrapins from Barnstable Harbor. By the 1900s, terrapin was a featured item on the menus of the upscale restaurants in large cities on the east coast, such as Delmonico's in New York and Haussner's in Baltimore. The herpetologist Roger Conant referred to the terrapin as "the most celebrated of American reptiles." The terrapin undoubtedly earned this epithet because of its palate-pleasing properties.

Not only was the terrapin a regional specialty, it was also exported to Paris, Berlin, and even South America. For those of us who have never tasted terrapins or turtle soup, it might be difficult to imagine why they were such an esteemed and sought after food. By all accounts, they were described as delicious. Coker (1920) wrote, "We may be sure indeed that the present preeminent position of the diamond-back terrapin among costly meat foods is based upon sincere gustatory discrimination and that its savory presence is approached with no other sentiments than those which become the highest gastronomic observance." (p. 171) Furthermore, "the diamond-back terrapin must have an inherent flavor that is held to justify the price at which it is purchased" (p. 185).

There are differing opinions with respect to preparing terrapins for the dinner table. A favorite Maryland recipe calls for placing a "count" (referring to a terrapin that had a plastron of at least seven inches in length) alive on its back in a stove, roasting it until the bottom shell could be easily detached, removing the gall bladder and then adding a little butter, salt, and a glass of sherry or Madeira. The terrapin was subsequently eaten right out of the shell.

Other recipes describe the preparation of terrapin soups and stews. All such recipes called for a heavy dose of sweet wine. From *The Household Cyclopedia of General Information*, published in 1881, we have the following recipe for terrapin:

> Plunge them into boiling water until they are dead, take them out, pull off the outer skin and toe nails, wash them in warm water and boil them with a teaspoonful of salt to each middling sized terrapin till you can pinch the flesh from off the bone of the leg, turn them out of the shell into a dish, remove the sand-bag and gall, add the yolks of 2 eggs, cut up your meat, season pretty high with equal parts of black and cayenne pepper and salt. Put all into your saucepan with the liquor they have given out in cutting up, but not a drop of water, add $1/4$ of a pound of butter with a gill of Madeira to every 2 middle sized terrapins; simmer gently until tender, closely covered, thicken with flour and serve hot.

A fancier version of terrapin stew can be found in *Delmonico's Recipes from a Gilded Age*, "A 1894 Thanksgiving Terrapin, à la Gastronome From the Table," by Alessandro Filippini.

> Take live terrapin, and blanch them in boiling water for two minutes. Remove the skin from the feet, and put them back to cool with some salt in the saucepan until they feel soft to the touch: then put them aside to cool. Remove the carcass, cut it in medium-sized pieces, removing the entrails, being careful not to break the gall-bag.
>
> Put the pieces in a small saucepan, adding two teaspoonfuls of pepper, a little nutmeg, according to the quantity, a tablespoonful of salt, and a glassful of Madeira wine. Cook for five minutes, and put it away in the ice box for further use. Put in a saucepan one pint of Espagnole sauce and a half pint of consommé. Add a good bouquet, one tablespoonful of Parisian sauce, a very little red pepper, the same of nutmeg, and half a glassful of Madeira wine. Boil for twenty minutes, being careful to remove the fat, if any; add half a pint of terrapin and boil for ten minutes longer. Then serve with six slices of lemon, always removing the bouquet.

Terrapin soup was a favorite White House lunch course during the presidency of William Howard Taft, a politician who enjoyed eating and had the waistline to prove it. The soup was also served during state dinners, when Mrs. Taft would hire a cook to prepare it. Mrs. Taft paid the cook $5.00 for this special effort. The Taft White House recipe can be found in *The President's Cookbook* (Cannon and Brooks, 1968).

### Taft Terrapin Soup

Brown 4 pounds veal knuckle in just enough fat or shortening to prevent burning. When it is a good crusty brown, add 2 sliced onions, 2 carrots, cut in half, 2 stalks celery halved, 3 cups tomatoes, preferably fresh, 1 bay leaf, 1/4 teaspoon marjoram, 1/4 teaspoon thyme, salt and pepper to taste and 3 quarts water. Simmer over low fire for approximately 3 hours. At that time, cut the meat from one turtle into 1-inch cubes and simmer it gently for 15 minutes in 1 cup sherry. Then strain the broth from the veal mixture and add it to the turtle meat. Mince 1 hard-boiled egg very fine and add to mixture. Simmer a few minutes and serve hot with slices of lemon floating on top. (If you prefer a thicker soup, blend in a little flour mixed with and equal amount of melted butter just before serving.) Serves 10 to 12.

At the insistence of President Taft, champagne was always served with the terrapin soup.

Of the seven subspecies, the northern diamondbacks were described as the most flavorful. They were known in the trade as "Chesapeakes," "Delawares," or "Delaware Bays," and "Long Island Terrapins." It was claimed that merchants could distinguish terrapins from different regions and were particularly vigilant about preventing the less desirable southern terrapins from being mixed in with their stock. However, it seems certain that southern terrapins infiltrated the commercial market. As Chesapeake terrapins became more difficult to find, wholesalers replenished their stocks with specimens from North Carolina. They were often mixed with Chesapeakes when they were shipped to northern markets. It is also certain that South Carolina terrapins made their way into the market in the same manner and were sold as Chesapeakes. Turtle meat was so popular that various freshwater species of *Graptemys* and *Pseudemys* (map and painted turtles) were harvested to meet the demand.

In the late 1800s and early 1900s, the economic importance of the terrapin spurred efforts to cultivate this turtle and led to the development of terrapin farming operations. Some of the earliest facilities were located in Charleston, South Carolina, and in the Maryland towns of Crisfield and Lloyds. The

farms were called pounds or crawls and consisted of a variety of empoundments or pens within which were various pseudo-natural areas containing
sandy ground above water, grassy areas, muddy tidal marsh, and muddy areas
covered by a few feet of water. These facilities were designed to answer several
important questions:

> 1. Is present legislation for the protection of this form based on sat-
> isfactory knowledge of the habits of the terrapin?
> 2. Can anything further be done by either the State or National
> Government toward checking the extermination of the terrapin?
> 3. Is it practicable to breed and grow the terrapin as a private enter-
> prise, as the Japanese do so successfully with their soft-shell snapping
> turtle, *Trionyx japonicus*, Schlegel? (Coker, 1906, p. 9)

Some scientists maintained that terrapin cultivation would be easier than
raising poultry. The U.S. Bureau of Fisheries expanded upon initial efforts to
farm-raise terrapins and asked for an appropriation from Congress to hire a
terrapin cultivator and set up a research and artificial propagation facility in
Beaufort, North Carolina. Pens were constructed and terrapins were reared
from eggs. The rationale for the setup of the empoundments was to provide a
number of separate areas that would substitute for natural environments.
Pounds contained areas for adults to feed, sandy spots for females to lay eggs
and separate spaces for young terrapins so they wouldn't be trampled by the
adults. The spaces allowed about 5 square feet (0.46 square meters) per adult
and 0.5 square feet (0.046 square meters) for each young terrapin. The facility at Beaufort also contained a hothouse, a heated building where a portion
of the hatchlings were raised for their first years. Fencing around the pounds
was designed not only to prevent terrapins from escaping, but also to prevent
poaching. In a summary of terrapin cultivation techniques, Hildebrand and
Hatsel (1926) warned, "It must be remembered, however, that so valuable and
easily marketed an animal as a diamond-back terrapin is a temptation that is
hard for a poacher to resist. It will be safest therefore to add a few strands of
barbed wire to the top of the inclosing walls and to enclose the pen by a
barbed-wire fence set back some 20 or 25 feet. The latter is particularly desirable, as it will not only make depredations difficult but will prevent inquisitive
visitors from approaching the pen" (p. 7).

Some feared that cultivation would allow terrapins to become as common
as potatoes and that they would lose their epicurean status. The slow growth
of terrapins, the necessity to protect them from predators, and the need to

provide large pens with suitable growth conditions made for a very labor-intensive venture. Although the facility was hatching 15,000 to 20,000 eggs per year, these attempts never achieved commercial success, even though the early analysis of the Beaufort facility suggested that terrapins are easy to care for, inexpensive to cultivate, and that one caretaker would be able to take responsibility for 50,000 turtles. By 1918, when multiple factors contributed to the decline in the terrapin soup fad, the Beaufort facility was closed.

Although edible, other subspecies were not harvested as heavily as the northern variety. The flesh of the Florida diamondback terrapin has been described as "inferior" (Pope, 1946), but Texas terrapins were readily consumed. In the mid 1800s diamondback terrapins were abundant along the coast of Texas. They were easy to net; many were caught and sold to local seafood markets or purchased by hotels in Galveston and Houston.

During the heyday of the terrapin soup era, there were several initiatives to hybridize farm-raised terrapins. In the herpetology classic *Turtles of the United States and Canada,* Clifford H. Pope described unsuccessful efforts to hybridize the large Texas diamondback from the Gulf of Mexico to improve the flavor, and hence the market value, of the Texas turtle. The hybrids grew slowly and matured later than the parent subspecies (Pope, 1946).

Although beginning to decline in popularity by the 1920s, terrapin was still featured at many gala dinner parties. Terrapin soup was the first course served at the inaugural Academy Awards banquet in Hollywood in 1929. This course was followed by jumbo squab Perigaux, lobster Eugenia, L.A. salad, and fruit supreme.

Terrapin dishes were not restricted to fancy restaurant fare. Terrapin also made its way into the everyday household, as evidenced by the inclusion of terrapin soup recipes in family cookbooks, such as the culinary classic *Joy of Cooking,* first published in 1931. Recipes for terrapin can also be found in the 1975 edition, in which the authors suggest that the diamondback terrapin is the "choicest of all turtle meat." The authors (Rombauer et al., 1975) described the preparation of terrapin (not a routine kitchen task):

> Sectioning it for cooking is an irksome job, even if you overcome the worst of the opposition—as old hands are wont to do when working with snappers—by instantly chopping off the head.
>
> Before preparation, however, it is advisable to rid turtles of wastes and pollutants. Put them in a deep open box, with well-secured screening on top; give them a dish of water; and feed them for a week or so on 3 or 4 small handouts of ground meat. (p. 393)

A detailed recipe was provided in which most of the turtle, including eggs, if present, is incorporated into the dish (Rombauer et al., 1975). In the editions of *Joy of Cooking* that are now generally available, there is no mention of turtles or terrapins.

Curious about the inclusion of terrapin-based recipes in contemporary cooking, I perused the cookbooks in my own collection. I found two recipes for terrapin soup in one of my favorite cookbooks, *Talk About Good!*, first published in 1967 by the Junior League of Lafayette, Louisiana. In true Creole/ Cajun style, the standard recipe is modified by the addition of Tabasco sauce. Most modern cookbooks have no mention of turtle preparation, or terrapin recipes. These dishes have all but disappeared from popular American cuisine.

Eventually, the numbers of terrapins declined throughout their entire range. Some local populations may have been entirely extirpated. After many decades of over-harvesting, the diamondback terrapin was in short supply. By the late 1800s, Chesapeake watermen noticed the declining numbers, but the high price of terrapin stimulated the continued harvest. It was not long before terrapins became increasingly difficult to find, and by the early 1900s, Coker (1920) remarked that "the majority of the diamond-back terrapins brought to market are taken more or less by fishermen pursuing other manner of prey" (p. 174) and "there can be no danger of inadequacy of food supply for the small remnant of terrapins that survives in the present day" (p. 173).

On Long Island, New York, large holding pens were constructed to supply terrapins to New York City. By 1916, local naturalists noted the decline of terrapin in Long Island waters. Terrapin sightings were limited to those specimens that were contained in chicken wire pens, awaiting a visit from a wholesale dealer. By the mid 1930s, Long Island terrapin were thought to be extinct.

Accompanying the decline in the terrapin population, the popularity of terrapin dishes began to wane. By the 1920s, a combination of factors most likely contributed to the shift away from terrapins as a high-status food item. Prohibition made it difficult to have access to the "spiritual condiments" such as wine that were needed for the preparation of terrapin soups and stews. The decline in the terrapin fishery as a result of overharvesting and the enactment of state regulations to protect the species made it more difficult for restaurants to procure terrapin. Wartime rationing, the Great Depression and subsequent changes in popular culture may have also been responsible for the decline in the demand for terrapin as a food item. When families could no longer afford to hire servants and cooks, the labor-intensive preparations to make terrapin soups and stews gave way to less expensive and more convenient dishes.

Some Gulf Coast fishermen began to consider the once exalted terrapin to be a sign of bad luck. They called it the "wind turtle" and believed that capturing one would cause ominous squalls and result in perilous voyages (Rudloe, 1979).

## Commercial Harvest

Although terrapin has been out of favor as a popular U.S. dish for over fifty years, there is still a demand for terrapin in ethnic markets. On a stroll through New York City's Chinatown on a February morning in 2002, I saw several fish markets displaying large plastic buckets filled with lethargic diamondback terrapins. On a subsequent visit to New York City in early April, 2005, I observed a 30-gallon garbage pail full of females as well as smaller males in a Mott Street fish market. I was unceremoniously escorted from the fish store when I took out my camera and tried to photograph the captive turtles. From my vantage point on the sidewalk, I noticed that the terrapins were sold off in less than thirty minutes. These reptiles are presumably sold for home use. Terrapins are also in demand in Asian markets because their shells can be used to predict the future and guide major decisions. Chinatown restaurants have long served traditional dishes prepared with terrapin, but these dishes are not usually seen on English versions of menus (Garber, 1986). It has been estimated that 10,000 terrapins are sold each summer in New York City, with single turtles sold by the pound and retailing for more than $20 each.

When I visited Chesapeake Bay in late April 2004, I was witness to the result of the spring season commercial harvest. I traveled east from Annapolis, over the majestic Chesapeake Bay Bridge, to Kent Island on the Eastern Shore. At the Horsehead Wetlands Center, Marguerite Whilden was temporary caretaker to more than a thousand terrapins. All but a dozen or so were large females. Using money raised by private donors, Whilden had purchased the terrapins from watermen for $4.00 each to prevent their slaughter. Where does one keep so many terrapins? They were everywhere. A garage was filled with shedding pens, large rectangular bins where molting blue crab are kept until they shed their hard shells and can be sold as "softshells." The pens were converted into terrapin holding tanks, and were very crowded (plate 2). Whilden checked on them every day and changed the water frequently. A cottage on the site, ravaged by Hurricane Isabel in September 2003, was temporary home to more terrapins. They were crawling all over the place and heaping themselves into huge, rocklike mounds in the corners of the rooms. It was not clear how many of the terrapins were specifically harvested and how many were by-catch, caught in fyke nets by shad fishermen. The turtles were packaged, twenty to a box, when Whilden loaded them into the back of her Jeep and transported them to the center. As she carefully unpacked them, it was apparent that some of them were in troubled health. Others were dead or dying. Whilden's intent was to tag and then release them when the official season ended on May 1. The problem that she faced was that she did not know the site of origin for these animals; they may have come from vastly different areas of the bay. All studies indicate that terrapins are a nonmigrating species and remain in the same area year after year. It is not clear how Whilden's rescue efforts will impact local terrapin communities in the bay. For now, these turtles will be released in areas where there is good feeding and nesting habitat and little in the way of crab pots and other fishing gear that can drown the turtles. Before their release, Whilden will attach small metal tags to the carapace of each turtle so there is a potential to follow their distribution over the coming years. She hopes that if the tagged terrapins are caught by watermen or researchers, their recapture will be a source of valuable information in the years ahead.

Some scientists have been critical of Whilden's strategy of buying terrapins from wholesalers. They fear that her activities are increasing the market demand for terrapins and will thus stir up the terrapin fishery. Some fear that terrapins will be released into areas where they will not be able to survive. It can also be problematic if they do survive because they may harbor pathogens or have a different genetic background from the resident population. The potential to alter the genetic structure of terrapins that have adapted to a particular region of the bay could be deleterious for a population.

Instead of buying and releasing terrapins, Whilden would prefer that the entire terrapin fishery be banned or that a moratorium on commercial harvest be mandated until a proper population assessment can be performed. She is willing to raise private funds to compensate fishermen as well as wholesalers for any profit they make in the terrapin fishery. She is promoting this idea and is ready to implement such a program if it ever gains favor.

With new Asian markets for a variety of turtle species, the sleeping commercial terrapin fishery may be waking up. For example, harvest records in North Carolina indicate that 460 turtles were captured in 2000. This number rose dramatically to 23,000 turtles in 2003. Although terrapins were not the most predominant species in the harvest, the high demand for turtles, combined with additional pressures, does not bode well for the mid-Atlantic terrapin population.

## Natural Predators

### RACCOONS

The many predators of diamondback terrapin eggs, hatchlings, and juveniles, including ghost crabs, raccoons, eagles, gulls, and rats, described in chapter 4, can be expected to take a toll on diamondback terrapin colonies. Most likely, there are also aquatic predators that fed on small terrapins. In the southern part of their range, the American crocodile (*Crocodylus acutus*) and the American alligator (*Alligator mississippiensis*) have been known to munch on adults, shells and all. Sharks may even find terrapins that venture into deeper waters. These threats are not unique to *Malaclemys terrapin*. Extensive predation on eggs, hatchlings, and juvenile turtles of many species is quite common. But adult turtles are armored by carapace and plastron; aquatic turtles are excellent swimmers. Thus, adults have few natural predators. There have been dramatic photos of adult diamondback terrapins clutched in the jaws of giant alligators, and some reports of shells of smaller terrapins, mostly adult males and some juveniles, in eagle nests in Florida Bay. Raccoons (*Procyon lotor*) are the only significant nonhuman predators of adult diamondbacks throughout their range.

Seigel (1980d) observed a raccoon attack on an adult female diamondback at the Merritt Island Wildlife Refuge. The female had apparently been on a nesting run. The raccoon broke the turtle's neck and was gutting the terrapin through a hole where the hind leg had been severed. On closer inspection, 24 other freshly killed terrapins, mostly adult females, were found along a 0.5 kilometer (547 yard) stretch of dike road surrounding the lagoon. The finding of old decomposed terrapin shells in the same area caused Seigel to speculate that perhaps up to 10 percent of the adult females in the colony were killed by raccoons from 1977 to 1978, clearly a significant dent in the reproducing members of this population. Seigel attributed the predation to an increase in the numbers of raccoons after the destruction of the salt marsh by the mosquito control dikes that were built in 1958 and the increased use of the dike roads for nesting by terrapins. Both factors led to a situation in which two species experienced increased contact as a result of alteration of the habitat by humans. When Seigel revisited this study site ten years later, terrapins could not be found (Seigel, 1993), and in all likelihood the local population has been extirpated.

Raccoon predation on adult diamondback terrapins was also frequently observed in Jamaica Bay Wildlife Refuge when raccoons first appeared in the refuge. There are gruesome accounts of eviscerated adult female diamondback terrapins, caught by raccoons during attempts to lay eggs (Feinberg and Burke, 2003). Curiously, raccoons in the refuge no longer prey on adult females. They have learned to be patient and wait for the eggs. In this way, raccoons expend much less energy to utilize a food source. An added bonus to the raccoons is the constant supply of eggs if the females are not killed.

Raccoons remain a significant diamondback terrapin predator, but the target for most raccoons is the nest. Many eggs and hatchlings are lost each year to raccoons (plate 7; fig. 3.7). A detailed examination of this problem is presented in chapter 3.

### MOLLUSKS

Occasionally, mollusks are found attached to diamondback terrapins (fig. 5.3). We can ask whether these epifauna are hitchhikers, causing no real harm to the turtles, or predators, inflicting damage and contributing to the mortality of individuals.

Barnacles and oysters can attach to the shell of diamondback terrapins much as they do to rocks, shells, cement and other hard substrates. Although the mollusks are seen on shells, the occurrence is not universal, nor very common. The first documentation of so-called "barnacle fouling" of diamondback terrapins was a report of a specimen collected in 1962 in Florida. This terrapin, which died shortly after capture, har-

*Fig. 5.3. Oyster on rear portion of female terrapin carapace, Wellfleet, Massachusetts.*

bored seven oysters (*Crassostrea virginica*) of different sizes and two types of barnacles, *Chelonibia patula* and *Balanus improvisus,* as well as the gastropod *Crepidulla plana,* commonly known as the slipper shell (Jackson and Ross, 1971; Ross and Jackson, 1972).

Perhaps mollusks are not commonly found on the shells of diamondback terrapins because of periodic desiccation when terrapins bask at the water's surface. Perhaps a terrapin carapace or plastron is not the ideal substrate for attachment of oyster and barnacle spat. Perhaps the terrapin that harbors the mollusk will move in areas with wide thermal and salinity variation that mollusks cannot tolerate. Perhaps the shedding of old keratin on scutes, a process known as ecdysis, removes any molluscan larval forms that have recently attached. Whatever the reason, fouling of terrapin shells by mollusks is an occasional but not a very common occurrence. Even rarer than barnacles on a terrapin carapace was a report of the bivalve *Brachidondes exusttus* growing within a vacant barnacle shell on the carapace of a diamondback terrapin (Jackson et al., 1973).

Attachment of barnacles and oysters to terrapin shells (carapace, plastron, and bridge), as well as to skin on the head and limbs, can create a situation in which the terrapin cannot properly move. Hydrodynamic drag, caused by shell adhesions, can interfere with swimming. There was a report of a terrapin so heavily encrusted with oysters on its rear carapacial scutes that it could only swim vertically. The terrapin was otherwise healthy and caused one observer to describe it as "the making for an oyster stew and turtle soup all at the same time" (Allen and Neill, 1952). It is also possible that an infestation of mollusks on the shells of terrapins can interfere with mating and copulation. A heavy coating of barnacles or oysters on the plastron of the male or the carapace of the female can inhibit successful reproduction.

Barnacles may be more than harmless hitchhikers or commensals (organisms that live together without causing each other harm). In a central Florida study conducted in the late 1970s, 76 percent of the 125 terrapins examined were infested with barnacles. Three species of barnacles were represented: *Balanus eburneus, Chelonibia manati,* and *Chelonibia testudinaria.* Barnacles were found most commonly on the carapace, but also on the plastron, bridge, and, on rare occurrences, the head or limbs (Seigel, 1983). This type of extra baggage may slow down terrapins and make them more susceptible to predators. In addition, barnacles may cause physical damage by eroding the shell beneath the area of attachment.

## *Natural Events: Unknown Causes*

On Cape Cod, it is common to find several dead terrapins each spring. The dead animals may represent those that have not survived the stress of hibernation or have died from injuries or natural causes. The number of dead terrapins rarely exceeds a dozen. The death of nearly 100 terrapins during the winter of 2000 remains unexplained. Washed atop the brown, stunted stalks of *Spartina patens*, terrapin carcasses dotted the marsh. Nothing like this had ever been seen. Examination of the carnage and assessment of relative decomposition suggested that the deaths were most likely the result of a late autumn event. The remains of terrapins, young, old, male, and female, yielded no clues to explain their demise. Theories to account for the massive die-off include an unknown terrapin parasite or disturbance of hibernacula by ice scouring, movement of boat moorings or dragging on the bottom of creeks by shellfishermen. There was no obvious damage to shells that would suggest physical injury, and the decomposed bodies were not likely to yield information about a bacterial, viral, or protozoan invader. Another possible explanation for these mysterious deaths was drowning. Perhaps the terrapins drowned as they sought calm shallows for hibernation and became caught under submerged obstacles such as plastic netting and other types of gear used by local aquaculturists. Those who walk in the marsh adjacent to shellfish beds are very familiar with the abandoned gear that washes off intertidal mud flats, particularly during winter storms. I have removed truckloads of plastic nets and PVC (polyvinyl chloride) pipe from marshes each spring. Quite a bit of it can still be used, so I recycle it back to my friends who farm oysters and quahogs in Wellfleet. Since the time of the massive terrapin kills, volunteers have been organized to remove shellfish gear that washes into creeks within the marsh, and there has not been another similar spike in terrapin mortality.

Hurricanes represent natural events that have the potential to impact terrapin populations. Tidal surges and habitat alterations caused by hurricanes do not seem to bother diamondback terrapins. In the few cases where hurricane effects have been studied, there has been no significant decline or dispersal of terrapins (Gibbons et al., 2001; Miller, 2001). Even fragile terrapin eggs and hatchlings in the nest can survive brief inundation by storm surges. While walking along the banks of the Chesapeake during springtime in a terrapin nesting area that had been under 2.4 meters (8 feet) of water after Hurricane Isabel the previous fall, Marge Whilden and I found a tiny hatchling that had either overwintered in its nest or hibernated in the nesting area.

## *Road Mortality*

A curious method to conduct an inventory of species in a certain area when there may not be enough personnel for conventional field studies is to use road kill surveys. It was just such a survey in 1994 that indicated that diamondback terrapins inhabit Guana River State Park in Florida (Charest, 1994). A dead terrapin on one of the park roadways must mean that there are terrapins in the park. Many species of turtles emerge as statistics in road kill surveys, especially if the surveys are conducted during nesting season or during times when hatchlings have emerged and are seeking water or cover. Terrapins are among the species that are occasionally found beneath the wheels of motor vehicles.

In southern New Jersey, the barrier beaches that once served as nesting grounds for diamondback terrapins have been radically modified. Sand dunes were leveled to accommodate the development of resort communities along the coast. Although the marshes of Cape May are now protected from development, the nearby waterways are heavily used, especially during the summer.

*Fig. 5.4. Turtle crossing sign on roadside tree near nesting area in Wellfleet, Massachusetts.*

Commercial and recreational uses include fishing, crabbing, swimming, boating, and water/jet skiing. Female terrapins emerge from these busy waterways and are faced with the challenge of finding suitable substrate and location to lay eggs. The only accessible areas are primarily on the unpaved shoulders of roads and causeways. Since 1989, the Wetlands Institute in Stone Harbor has embarked on a long-term research and conservation project to determine the extent of the road kill problem and to devise measures to prevent this type of mortality.

On a typical summer's

day or night during nesting season, diamondback terrapin fieldwork may consist of monitoring a 16-kilometer (10.5-mile) stretch of pavement and scooping up road-killed females who were on a mission to deposit their eggs. From the carnage, it is sometimes possible to carefully salvage eggs by dissecting them from the dead female's oviduct. The Wetlands Institute has set up an exhibit in which the seasonal tally of road-killed females is prominently displayed. Despite signage that directs motorists to slow down for turtles, close to 500 females per year are the victims of automobile accidents. From 1989 to 1995, 4,020 terrapin road kills were reported (Wood and Herlands, 1995). Diamondback terrapins have been killed on roadways throughout their range, but the problem remains chronic and extensive in New Jersey.

I have found diamondback terrapins killed even on dirt roads where homeowners are very alert to the possibility of a wandering female during nesting season. The terrapins are sometimes very difficult to see, especially when they are not moving, because they are often covered with sand and blend in with the roadway. Many of them actually select the sand road for their nesting site. Despite signs that alert motorists to a turtle crossing (fig. 5.4), tourists, utility companies, and delivery vehicles may not be aware of nesting terrapins and may not be looking out for them. When a tire-track nest appears to be doomed, it can be relocated to safer territory nearby. Due to the nature of hatchling development, we must always take care to preserve the depth and solar exposure of the nest and the orientation of the eggs during the relocation process.

If a female is struck by a car but is still alive, prompt veterinary care may save her life and allow her to reproduce for years to come. If internal organs are not damaged, her shell can be wired or fiberglassed until it heals (fig. 1.7). Recovery may take a year or more, so it is common to keep the female in captivity until she has recovered.

## By-Catch

The tasty blue crab, *Callinectus sapidus*, shares much of the range of diamondback terrapins. Although the diamondback is no longer a popular food item, the blue crab still commands a large commercial and recreational fishery. Crab traps, or pots as they are commonly called, are large metal wire boxes that are fitted with entrance holes on more than one side. The typical Maryland-design crab traps are 24 inches by 24 inches (61 centimeters by 61 centimeters) and 21 inches (53 centimeters) deep. They have entrance funnels, 11 to 12 centimeters (4.3 to 4.6 inches) wide, at the base of each side. The pots are fixed with bait to attract the crabs. Mullet are commonly used as bait in the South; menhaden, commonly called "bunker," are popular bait in the North. When crabs enter the traps, they are unable to leave. The design is ideal for catching crabs, and the pots can be placed in the water and checked as often as needed or as often as required by permit or license. Unfortunately, the pots also attract diamondback terrapins, some of which may be attracted by the bait, and others make their way into unbaited traps. Depending on the size of the openings, terrapins will be able to enter the pots but may not be able to navigate an exit. If pots are not checked frequently enough, terrapins will drown.

The incidence of terrapin by-catch in some locations may follow seasonal trends. For example, terrapin capture in pots in South Carolina is highest in April and May and may be associated with post-hibernation foraging and mating activity in the areas around subtidal mudflats (Bishop, 1983). In a 1979 to 1981 study in Charleston County, South Carolina, Bishop (1983) employed various types of crab pots to assess the extent of terrapin by-catch by the local commercial crabbing industry. Over a three-year period, 281 diamondback terrapins were caught in the traps, with an average ratio of males to females of 2.3:1. Large females were restricted from the pots due to the size of the trap entrance holes. A mortality of 10 percent was observed when the pots were checked every day. When Bishop extrapolated his data to the possible number of terrapin by-catch, he used the number of commercial crabbers (743) and the average number of pots that were fished by each crabber and estimated that 2,853 terrapins may be caught per day with a daily mortality of 285. This translates into a mean daily terrapin by-catch of 0.16 per baited crab pot during April and May. Although some of the mortality appeared to be due to predation of the trapped terrapins by blue crabs, most deaths were from drowning. When the study was completed in the 1980s it was thought that the level of by-catch would not have a significant impact on the Charleston terrapin population. Now that we are more aware of the local population numbers and age structure, it seems that even these low rates of by-catch can be tremendously harmful to local terrapin clusters. Based on the number of permits issued, it can be estimated that over 20,000 commercial and 35,000 recreational crab pots are fished annually in South Carolina.

South Carolina is not the only state where crabs and terrapins are in potential conflict. The blue crab fishery has tremendous economic impact on the entire mid-Atlantic and Gulf regions. Data gathered in New Jersey by Roger Wood and Roz Herlands (1995) are troubling for terrapins. Using 1993 as a "typical" example, 63,280 licensed commercial crab traps were used along the New Jersey coast. In addition, 4,865 recreational licenses were issued and each recreational crabber was allowed to fish two pots. In areas in which terrapin are abundant, field experiments indicated that about 5.5 terrapins are caught per 100 traps per day. Of the terrapins that end up in the crab traps, one-third of them drown. A quick calculation revealed that during a season in which there are five months of active crabbing,

63,289 traps x (5.5 terrapins/100 traps) = 3,480 terrapins.
3,480 terrapins caught x 1/3 drown = 1,160 terrapins drowned/day
5 months of active crabbing (153 days) 1,160 terrapins drowned (per day) = 177,480 drowned terrapins. (Wood and Herlands, 1995, p. 256)

Even if only a fraction of the traps are placed in waters where terrapins are abundant and likely to enter them, a significant impact on the population may be predicted.

Commercial-type crab traps are so effective in catching terrapins that they have been used as tools by researchers to determine terrapin abundance and distribution. But a crab trap is not the only device that can inadvertently catch a terrapin. Along the Texas coast, terrapins can be accidentally snared in shrimp trawls, and in Chesapeake Bay, terrapins are found drowned in eel pots.

Although commercial crabbing is a serious threat to diamondback terrapins, we cannot overlook the contribution of recreational crabbers to accidental terrapin mortality. In a South Carolina creek, near Kiawah Island, a well-characterized population of diamondback terrapins was used as a baseline to study the impact of recreational crab pots on local terrapins. In a population of 168 to 299 terrapins, nineteen individuals were caught in recreational pots during 760 crab pot-days (the number of pots x the number of days). The pots caught mature males and immature females. The number of terrapins that were caught during the time period of the study represented 6 to 11 percent of the total creek population (Hoyle and Gibbons, 2000). It is not difficult to see how recreational crabbing can have a severe impact on local terrapin populations.

Crabbing may take a differential toll on terrapins depending on whether pots are set in deep water or in more shallow creeks and channels, close to shore. Roosenburg et al. (1999) showed that different age classes and different sexes of terrapins may utilize different habitats. In certain Chesapeake Bay locations, adult females are commonly found in deeper waters, while adult males and juveniles of both sexes forage closer to shore. Depending on placement of crab pots, different segments of the terrapin population may be more susceptible to crab pot mortality.

In the Patuxent River, Roosenburg et al. (1997) estimated terrapin bycatch rates of 0.17 terrapins per pot per day in shallow water with a 3:2 male bias. Due to sexual size dimorphism, larger, mature females were immune from capture because they can't fit into the crab pot openings. In contrast, mature males are at risk for their entire lifetime. With a good population study as a baseline, it was estimated that crab pots could eliminate 15 to 78 percent of the local population within a year. Even if the lower estimate proves more accurate, the eventual effect on the population would be devastating and would cause extirpation within a few years. From a conservation perspective, one of the more disturbing aspects of this study, as well as the study by Hoyle and Gibbons (2000) described earlier, is that sometimes recreational crabbers may have much more of an impact on terrapin populations than the commercial crab fishery. Recreational crabbers usually set their pots in shallower water, precisely where males and juveniles are found.

The longer the crab pots are unattended, the greater is the possibility of death by drowning. When crabbers take a day off from checking their gear, the likelihood that they will find dead terrapins increases. Pots that become abandoned or lost are also a serious problem. These lost pots, known as ghost pots, may have become loosened from their original location and are no longer monitored. Sometimes they are simply abandoned or forgotten by vacationers. The pots may shift in location due to tides, currents, and wave action. The ones that wash into shallow creeks are more likely to catch terrapins than those in deeper, open water.

Terrapins sometimes play "follow the leader" when it comes to crab pots. It has been observed that after one terrapin enters a pot, others are sure to follow. When terrapins are caught in unbaited pots, they are not usually alone. Mixed-sex captures are common. Perhaps when males are in pursuit of a female and she enters a crab pot, the males will follow her. This type of activity contributes to the demise of the reproducing members of the population.

To assess the impact of crab pot design on terrapin mortality in New Jersey waters, Roger Wood conducted a study in which several types of crab pots were employed. In New Jersey and points north, crab pots are called crab traps. He compared crab and terrapin capture in floating traps to that of unmodified traps that routinely catch blue crabs, spider crabs, conchs, and fish, as well as adult male and subadult female terrapins. The floating traps were designed so that Styrofoam floats would keep the upper section of the trap above water, providing a breathing space for terrapins. Unfortunately, the floating traps were not very effective in catching crabs so it was not feasible to promote their use as a terrapin conservation strategy.

In 1992, Wood designed the first prototype device that could be placed on the entrance funnel to crab traps and had the potential to prevent terrapins from moving into them. Wood's goal was to reduce the aperture size so that it would be too small for most diamondback terrapins but remain large enough to catch crabs. The device was called a by-catch reduction apparatus, but the acronym, BRA, did not become very popular. Such devices are now called TEDs, Terrapin (or turtle) excluder devices, or BRDs, by-catch reduction devices (fig. 5.5).

Wood and his colleagues engineered several types of TEDs, ranging from simple horizontal wires across the entrance funnels to rectangular wire or plastic frames of various sizes. The first TED prototype was not effective in field trials. Terrapins of all sizes made their way into the traps. A second design featured a 5 x 10 centimeters (2 x 4 inches) rectangular frame, constructed from wire coat hanger and attached to the funnel entrance. The design was promising; it did not reduce crab capture but was somewhat effective in decreasing terrapin capture. Only males and juvenile terrapins were able to enter the traps. Large females were excluded. When the rectangular device was made a bit smaller and was reduced to 4 x 8 centimeters (1.6 x 3.1 inches), all terrapins were excluded but crab captures decreased considerably. Clearly, this design would not be used by crabbers. Wood's group tested various size modifications of the basic rectangular design, always keeping in mind that if crabbers were to be convinced to use the excluders, crab catch could not

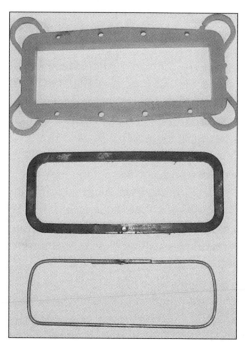

*Fig. 5.5. Sample TEDs/BRDs supplied by Roger Wood.*

diminish. After considerable testing, the 5 x 10 centimeters (2 x 4 inches) design was shown to be the most effective, even though some males and juvenile terrapins were still caught. Wood's group was encouraged by the fact that when the excluders were used, no reproductive-age females were being removed from the population by drowning in the traps. An added bonus of Wood's two-year study was that traps with excluders caught 9,675 market-size crabs, while the same number of unmodified traps, fished in the same locations, caught only 8,706 market-size crabs. The use of excluders netted an 11 percent increase in crab catch! When Wood continued the study for an additional season, traps fitted with excluders caught almost 40 percent more crabs than control traps (3,237 versus 2,167) (Wood, 1997). It seems curious that traps with excluders will yield more market-size crabs.

Roosenburg et al. (1997) prototyped a crab pot design for use in shallow water. The pot is anchored to the substrate, and the design allows for an air space that prevents terrapins from drowning. The design would work well in areas that are not subjected to large tidal variations. The trial study showed no difference in the modified versus regular crab pots in their ability to catch the most commercially valuable crabs. Thus, using the alternative traps would not compromise the catch of recreational crabbers. However, commercial crabbers often fish in deeper water than the design allows.

Wood's New Jersey study was not the only one that showed that TED/BRDs may actually increase revenue for crabbers. A similar TED/BRD study was conducted in Maryland. BRD designs were tested near the Patuxent River terrapin colony. The overarching purpose was to "balance the economic concerns of equipment cost, economic efficiency, revenue loss and the environmental concerns of diversity, sustainability and conservation" (Roosenburg and Green, 2000). When BRDs were used, researchers looked not only at the ability to exclude terrapins but also at the number, sizes, and types of crabs that were captured after employing various types of BRDs on crab pots. The 4.5 x 12 centimeters (1.78 x 4.7 inches) BRD was the most effective without impacting the crab catch, reducing terrapin by-catch by 82 percent (Roosenburg and Green, 2000).

In Louisiana, crab pots also have the potential to take a bite out of terrapin populations. A field trial, conducted by Guillory and Prejean (1998) from the Louisiana Department of Wildlife and Fisheries, tested the 5 x 10 centimeters (2 x 4 inches) TED designed by Roger Wood in three locations in the Terrebonne/Timbalier Bay estuary in Laforche and Terrebonne Parishes: Bay Blanc, Pointe au Chien Wildlife Management Area, and Bayou Blue. Their traps consisted of vinyl-coated wire mesh, 60.9 centimeters (24 inches) wide and deep and 36.8 centimeters (14.5 inches) high, with three entrance funnels. The total yield for control and TED-outfitted traps was 4, 145 blue crabs. Although no terrapins were caught during the study, there was a curious finding: more crabs were caught in traps with TEDs than in control traps. This was true for legal-size as well as smaller crabs.

For Joe Butler and George Heinrich's study of BRDs in eight Florida counties, a 12 X 4.5 centimeters (4.7 X 1.8 inches) device was employed. Field trials indicated that when pots with this design are compared to control pots, there is no difference in the number of crabs caught, the number of legal-size crabs caught, the size of crabs, or the ratio of male to female crabs. In assessing whether the devices decrease terrapin mortality, it was found that thirty-seven terrapins were caught in control pots, while only four were caught in traps fitted with BRDs (Butler and Heinrich, 2004).

From the TED/BRD studies that have been conducted in New Jersey, Maryland, northern Florida, and Louisiana, it seems clear that certain devices designed to prevent terrapin capture and drowning will not be a negative economic incentive for crabbers. At least the same numbers of crabs (sometimes more) are caught in pots outfitted with simple devices that prevent the entry of many terrapins. The devices most likely decrease the ability of crabs to leave the pots and would thus provide a bonus to crabbers. Because they are inexpensive, their installation would not be an economic burden for crabbers.

Although crab pots fitted with BRDs/TEDs decrease terrapin by-catch, some terrapins are still able to enter and become trapped. The vulnerable group consists of males and smaller terrapins of both sexes. The long-term effect of this skewed mortality may be a shift in the age and sex ratio in the populations. The terrapins that are not crab by-catch victims will be all the older, larger females.

## Pollution

### CHEMICAL AGENTS

Jamaica Bay, New York, has historically been one of the least pristine brackish water habits a terrapin could love. Surrounded by New York City landfills, shadowed by jumbo jets making their way to and from John F. Kennedy International airport, the area is not fit for swimming, and anglers might be considered foolhardy if they ate their catch. The fact that terrapins are abundant in Jamaica Bay defies explanation. Estuaries, marshes, lagoons, and other diamondback terrapin habitats such as Jamaica Bay are waterways that have the potential to collect chemical and microbial pollutants. Materials generated by natural biogeochemical cycles, agricultural and surface runoff, industrial waste, partially treated urban wastewater, and deposited airborne pollutants from industry and automobiles are all potential sources of toxic products. Yet diamondback terrapins have prospered in some of the most polluted waters on the Eastern Seaboard.

Pollution is sometimes difficult to trace. Although some pollutants are generated from point sources such as industry, dredge spoil and sewage, other sources are less clear. Pollutants may collect as a result of runoff from lawns and golf courses, dumps and roadways. A few studies have been conducted to assess the extent of exposure of diamondback terrapins to environmental pollutants and the ability of terrapins to accumulate pollutants in their tissues. The U.S. Geological Survey (USGS) has summarized reports about the levels of organochlorine pesticides, trace elements, and radioisotopes in tissues of diamondback terrapins from studies conducted in Georgia, New Jersey, and Florida and posted the findings on its web site (http://www.pwrc.usgs.gov/bioeco/terrapin.htm). In the few studies represented by the data, most pollutant levels in terrapin tissues were not remarkable and there were no apparent effects on the health of terrapins in the populations that were sampled. These observations must be carefully interpreted because no data were available on terrapin population status or health of individuals prior to their exposure to the toxic compounds that were used at or near terrapin habitats. In order to accurately assess the ecotoxological effects of specific compounds on diamondback terrapins, it is important that data be available about preexposure status of the populations.

As carnivores, adult terrapins sit at the higher trophic levels of the marsh food web. Shore birds that consume fish are also situated at the top of the marsh food web and have often been known to bioconcentrate pollutants such as mercury, found in fish tissue. Like some other contaminants, mercury makes its way into sediments, where microbes convert inorganic mercury to organic forms such as methylmercury, more easily taken up into animal tissues. When pollutants like mercury are subject to improper disposal and make their way into the marsh system, it is expected that birds as well as terrapins will be victims of bioaccumulation and thus will have high levels of the compounds in their tissues. Thus techniques are being developed to assess mercury biohazards using terrapins, along with birds, as top marsh consumers.

In a study conducted in Georgia near a superfund site contaminated with Arochlor 1268, a polychlorinated biphenyl (PCB) mixture used as a machinery lubricant, levels of the contaminant were lower in terrapin tissues than expected. It was lucky for the terrapins in this case that the pollutant appeared to have lower membrane permeability than expected and did not get into the animal's system as readily as predicted (Kannan et al., 1998).

## HEAVY METALS

In the case of heavy metals, a single study of terrapins in Barnegat Bay, New Jersey, measured lead, mercury, cadmium, chromium, manganese, arsenic, and selenium. Metal levels in muscle were lower than the limit imposed for commercial fisheries. However, analysis of tissues from eleven adult females indicated that most of the metals accumulated at higher levels in liver than muscle and that all metals were transferred to some extent from females to their eggs. A similar observation has been noted for freshwater and marine turtles. It was concluded from the study that metal accumulation in muscle would not be high enough to cause toxicity to anyone who consumed terrapin meat but could be problematic for consumers or scavengers who ate the liver of these turtles (Burger, 2002).

## OIL POLLUTION

Heavy metals may not represent the most serious current environmental pollutant that can harm diamondback terrapin populations. A more sporadic but very detrimental environmental impact may be caused by oil spills. Although major oil spills are dramatic and well publicized, they may not represent the bulk of oil contamination in coastal waters, much of which originates from routine industrial operations, seepage from production sites, industrial waste, and recreational boating. Much of this oil finds its way into our estuaries, where it is less likely to be dispersed by wave action and where it will quietly collect on the surface waters and beaches of coves and embayments. Many nonmigrating saltmarsh organisms and, in particular, sessile ones such as clams, mussels, and oysters are particularly vulnerable. When it comes to oil spills, diamondback terrapins are not lucky turtles. The most serious oil spills in the United States have occurred in terrapin territory, including the Florida spill in 1969 in Buzzards Bay, Massachusetts, the Bouchard 65 spill in 1974, also in Buzzards Bay, and the Exxon Refinery Spill in 1990 in Arthur Kill, New Jersey, all of which leaked number 2 fuel oil into coastal waters.

In the 21st-Century, there already have been two dramatic oil spills that have contaminated terrapin habitats. On April 7, 2000, an oil pipeline that ran 3 feet under the marsh at the mouth of Swanson's Creek, Maryland, cracked and began to leak fuel oil. The pipeline was used by the Potomac Electric and Power Company (PEPCO) and provided fuel for the Chalk Point Power Plant. An estimated 140,000 gallons of oil seeped into the marsh. Initial attempts to contain the oil with floating barriers were thwarted by a storm, resulting in the contamination of 40 miles of Patuxent River shoreline.

There were more than twenty-five scientific studies to assess the impact of the Chalk Point spill on the Chesapeake Bay ecosystem. A few of the studies were specifically geared to assess diamondback terrapin survival and the impact on reproduction of the species. After reviewing the studies, the Maryland Department of Natural Resources assessed the damage as follows:

- 76 Acres of wetlands had been contaminated with oil.
- 10 Acres of beaches had been oiled.
- 600+ Ruddy ducks and other birds had been killed.
- 122 Diamondback terrapins had been killed, and the future reproduction of this species had been reduced by 10 percent for the year following the spill.
- 376 Muskrats had died.
- 5,000+ Pounds of fish and shellfish had been killed.
- 4,000+ Pounds of creatures from benthic communities (the ecosystems on the bottom of the river) had been killed or reduced.

Aside from the toll on the ecosystem, the oil spill also had an effect on private property and the local economy; it also reduced boating, swimming, fishing, commercial, and recreational activities. The response to the spill by state and federal agencies, as well as by PEPCO, prevented wider ranging disaster. A large cleanup and restoration plan, costing millions of dollars, was put into place. The creation of six acres of tidal marsh next to Washington Creek and a one-acre beach habitat for diamondback terrapin nesting was also proposed.

Buzzards Bay is a busy shipping route for small tankers and barges that bring much needed heating oil to Boston and northern New England. It is estimated that 1.6 billion gallons of oil transit through Buzzards Bay and the Cape Cod Canal each year. Some of the oil does not make it to its final destination. Occasionally, a bit of it ends up in the bay. Grounded oil barges are the source of most of the oil, although in 1992 the oil leaked from a ruptured fuel tank of the *Queen Elizabeth II* cruise ship. Three years after the Chalk Point, Maryland, oil spill, the Bouchard Number 120 oil barge apparently deviated from the normal shipping lane and ran aground on Hen and Chicks Reef in Buzzards Bay. The barge may have trailed oil for 10 miles before being noticed by a tugboat captain. The slick was about 10 miles long and 2 miles wide and was partially dispersed by winds and seas. It was initially estimated that 14,000 gallons of number 6 fuel oil had spilled, but the figure was increased to 55,000 gallons in the days following the leak, and some Coast Guard reports put the level at about 100,000 gallons. This represented the largest oil spill in the area in thirty-five years. Despite the rapid action of emergency cleanup crews, and the placement of containment booms around the slick, over 90 miles of shoreline were contaminated. The Massachusetts towns that were affected included Bourne, Dartmouth, Fairhaven, Falmouth, Gosnold, Marion, Mattapoisett, New Bedford, Wareham, and Westport. Some of the neighboring Rhode Island coast was also oiled. Shellfish beds and fishing areas were closed. The most dramatic mortality of the spill was seen in the bird populations: There were casualties among twenty-nine species. On November 19, 2004, Bouchard Transportation Company agreed to pay a criminal settlement for the catastrophic oil spill caused by one of its barges. The settlement represented a plea bargain: Bouchard admitted guilt for the killing of birds in Buzzards Bay as a consequence of the spill. Although Bouchard has agreed to pay $10 million, actual long-term cleanup costs are projected to be ten times higher.

Buzzards Bay is historic habitat for diamondback terrapins but they had not been seen in the area for many decades. When a tiny turtle hatchling wandered into the Schaffer Oceanography Laboratory at Tabor Academy in Marion, Massachusetts, teacher Sue Nourse became intrigued. When Nourse found out it was a diamondback terrapin, she began to investigate whether the hatchling was an escaped pet or if terrapins were in the waters off Tabor's campus. When she discovered some depredated nests and had occasion to spot some terrapin heads in Sippican Harbor, she was able to confirm the presence of diamondback terrapins in the upper northwest reaches of Buzzards Bay. Since the population status of these terrapins is not known, the impact of the Bouchard 120 oil spill and other earlier spills on the Buzzards Bay terrapin population may never be revealed.

## Microbes

Microbes such as bacteria are present everywhere in the biospere. However, there are certain microbes that are typically found in the digestive tract of humans and other warm-blooded animals. These bacteria, which include *Escherichia coli* and *Salmonella* species, are referred to as coliforms, and certain species (fecal coliforms) are excreted by animals. Levels of fecal coliforms are used as indicators of water purity, and these microbes are monitored to determine whether we can drink the water (potable water quality), whether we can swim in the water (recreational water quality), and suitability of water for eating fish and harvesting filter feeders such as clams, oysters and scallops. Waters are closed when fecal coliforms reach a certain threshold, which varies depending on the intended use of the waterway.

In addition to human waste, fecal coliforms that are found in estuaries, marshes and other coastal habitats may originate from nearby farms that raise cows, pigs, or chickens. Waterfowl represent another major source of these bacteria and in some locations may contribute most of the load of fecal microbes in a habitat. There have been some attempts to culture fecal coliforms from diamondback terrapins. Cloacal and fecal sampling of terrapins has been conducted in Duval County, Florida. Coliforms were found in 80 percent of the fecal samples and in 51 percent of cloacal swabs. Microbes that were found include *E. coli, Citrobacter freundii, Klebsiella pneumoniae,* and *Enterobacter cloacae.* The levels of the microbes were lower than those that are expected for birds and mammals. In the same habitat where the diamondback terrapins were captured, fecal coliform levels in the water were within standards for water quality (Harwood et al., 1999). It is not clear how or if diamondback terrapins are contributing to the coliform levels found in the habitat. This bit of information may be important to know, since many of the terrapin habitats are also recreational and/or commercial shellfishing grounds. It is also not clear whether the terrapins are the primary source of their own coliforms or whether the turtles have been colonized by these microbes as a result of their introduction into the marshes by humans, birds, or other animals. So the question remains: Are fecal coliforms a threat to diamondback terrapins, or are terrapins contributing a significant amount of fecal coliforms to marshes and estuaries?

## *Phytoplankton and Macroalgae*

We do not typically think of microscopic algae or seaweed as harmful organisms, and for the most part this is a valid assessment. However, there are circumstances when the tiniest of algae can be deadly. The health of a marsh or estuarine ecosystem can be jeopardized in short-lived periods when harmful algae grow in abundance and produce toxins that are poisonous for fish, other animals, and even humans. Harmful algal blooms have been responsible for mass kills of wild and farm-raised fish, shellfish, and sea birds. In humans, their toxins my produce diarrhea, paralysis, neurotoxicity, and memory loss. Although most pigmented algae do not produce toxins, a few species are responsible for toxic "red tides."

The algae and their toxins concentrate in mollusks such as clams and oysters, which filter the deadly brew and then pass it up the food chain. The impact of harmful algal blooms on diamondback terrapins has not been reported, but as terrapins are mollusk eaters, there is a potential for the algal toxins to affect the health of the population. This is particularly troubling because the incidence of harmful algal blooms is on the rise. The cause of the blooms is also a mystery, although currents and global climate change have been implicated. Some researchers think that human activities may be responsible. For example, availability of nitrogen and phosphorus normally limits growth of phytoplankton. Pollution of coastal waters by nitrogen- and phosphorus-containing compounds as a result of agriculture, industry, or even lawn maintenance may cause an increased growth of the troubling organisms or trigger the production of their toxins.

A more long-term algal problem is the growth of large mats of seaweed, macroalgal blooms, a phenomenon that has also been increasing. Coastal development with resulting runoff has been identified as one source of the problem. The large mats of seaweed choke out sea grass beds that serve as nurseries for fish and may also be foraging grounds for diamondback terrapins. While tracking down the origin of a report in which terrapins were spotted in Mashpee on Cape Cod, I arrived at Waquoit Bay, only to find that this watershed, part of the National Marine Estuarine System, had been almost completely deprived of sea grass as a result of enormous growth of macroalgal mats. Although some of the upland areas looked ideal for diamondback terrapin nesting, the health of the bay and adjoining marshes had been severely compromised. I did not see any signs of diamondback terrapins and remain dubious that they could remain in such a degraded habitat.

## *Habitat Degradation*

Coastal areas have traditionally drawn settlers and have been hot spots for development. Humans have left their mark on all vistas of these varied ecosystems and have changed the ecological dynamics within salt marshes. Many terrapin habitats have become permanently fragmented, a condition that will further limit the already low levels of movement that terrapins display. Bulldozing dunes around coastal marshes to provide better water views is a blatant assault on terrapin nesting habitat. Other, smaller scale alterations of habitat occur when dune buggies and off-road vehicles (ORVs) compact sand, destroy vegetation, and disrupt the activities of animals. Some of the same perturbations occur on an even smaller scale as a result of foot traffic due to nature lovers, birders, hikers, and even researchers.

It may well be the alteration of the salt marshes in Merritt Island, Florida, that was responsible for the extirpation of diamondback terrapins by 1993. In the late 1970s, the Merritt Island terrapin population was studied by Seigel (1980a, 1980b, 1980c, 1980d). Although the population status could not be ascertained, adult terrapins were found in the marshes, and females were observed to nest along a dike road that was constructed in the 1950s. Over the years, much of the natural salt marsh was eliminated near the Merritt Island Wildlife Refuge. When Seigel revisited the area in 1993, there were no terrapins to be found (Seigel, 1993).

Human activities have greatly changed the number and extent of salt marshes all along the geographical range of diamondback terrapins. In some cases, wetlands have been filled; in other instances, the flow of water, the critical circulatory system of the marsh, has been altered. During colonial times, *Spartina patens*, salt-marsh hay, became a valuable commodity. Salt marshes were used as grazing areas, and salt-marsh hay was harvested for winter feeding of livestock. The hay was also used to stuff mattresses and as a material for insulation. Marshes were partially drained to promote the growth of *Spartina patens*, which prefers the drier regions of the marsh.

A greater impact on coastal marshes was inflicted by sometimes misguided attempts at mosquito control. Initially, many waterways were drained to prevent breeding of mosquitoes that could potentially transmit disease such as malaria. Tidal marshes were altered by filling or ditching. The ditches were narrow channels that promoted the flow of water out of the spongy marsh. The marsh substrate that was removed during ditching operations became an important source of fertilizer. Some of these ditches can still be seen in aerial photographs as dramatic grids across the face of coastal marshes. These efforts, which became public works projects, were also deemed important because they provided federal and state employment during the post-World War I era and the Great Depression.

When chlorinated pesticides such as dichlorodiphenyltrichlorethane (DDT) became available, they became a preferred component of mosquito control programs because they could be sprayed from airplanes. After malaria-carrying mosquitoes were "controlled" and coastal communities began to develop as resort areas, officials continued to ditch and drain marshes with the goal of taking aim at nuisance mosquitoes that made life miserable for inhabitants and vacationers during certain times of the year.

Along some shorelines, millponds were constructed by installing tide gates that retained water brought in on the high tide. Water was then let out through a narrow opening in which a waterwheel was mounted. The energy brought about by restriction of tidal flow was used to power the mills. In other areas, dikes were built to control or restrict tidal flow. The impairment in tidal action also caused salt marshes to be under water for prolonged periods, thus killing off much of the salt-marsh vegetation that only thrives with periodic drainage. Introduction of fresh water to formerly brackish areas diluted the salt content and thus "freshened" the water, promoting the growth of fresh-water grasses and plants such as the invasive marsh reed *Phragmites australis*. Marshes in South Carolina were circled with earthen barriers, known as dikes. By restricting tidal flow, the dikes created impoundments that were first used for rice culture and later maintained to attract waterfowl.

Salt marshes were considered to be nuisance areas, and thus many were filled to convert the land to more usable form. The fill was obtained from dredging projects that created navigation channels or deepened harbors. Some fill was composed basically of garbage and waste material. With marshes filled in, increased land for agriculture became available. Roadways could be built, train tracks could be laid, and development could proceed. The little tidal flow that remained was channeled into culverts or pipes that ran under roads or tracks. Much of Boston's upscale Back Bay neighborhood was a salt marsh near the mouth of the Charles River. When tidal flow was restricted by the Mill Dam and authorities considered the area to be a health hazard, the marsh was filled. Downtown Providence, Rhode Island, was called Great Salt Cove before it was filled. The human-initiated restructuring of these habitats was considered to be a sign of progress. Although difficult to estimate, the amount of salt-marsh alteration that has occurred is clearly extensive. Perhaps half of Atlantic coastal salt marshes have been significantly altered or eliminated as a result of human "improvements."

Our change in attitude about the marsh has come about slowly. Draining of wetlands was eventually shown to cause mosquito control efforts to back-fire. Mosquitoes can breed in small puddles, and no amount of ditching and draining can eliminate all standing water from a wetland. Furthermore, small fish that eat mosquito larvae need pools of water to swim, forage, and breed. Thus, ditching and draining may have the undesired effect of increasing mosquitoes. Small, mosquito-eating fish are preyed upon by larger fish and birds. Therefore, the impacts trickle through the entire marsh food web.

With the loss of historic salt-marsh habitats, diamondback terrapins have lost foraging, mating, basking, and hibernation territory. Disturbances in the water are not the only negative environmental impacts for terrapins. Females are losing nesting areas. Some nesting habitat disappears as a result of natural environmental forces. Nesting beaches, dunes, and marsh uplands are constantly reshaped by erosion and flooding due to the action of wind, waves, and currents. Generally, these forces work in cycles, at times depositing sand and sediments to build beaches, and at other times, eroding beaches. Such cycles usually occur gradually and over long periods of time. Sea-level rise, attributed to global warming, may also put diamondback nesting areas under water.

Human activities often accelerate the pace of shoreline alteration, directly as well as indirectly. Many marshes are sinking and may be completely under water in the near future. This phenomenon, known as subsidence, occurs when the pace of marsh substrate buildup, caused by sediment trapping and the compaction of dead marsh vegetation, does not keep up with rising water levels. This may be a consequence of the construction of dikes, which have a twofold effect on marsh systems. They restrict sediment flow from reaching the marsh below the dike and cause a rise in water level above the dike. The rising water level drowns marsh plants by preventing the periodic drying that is required to keep their roots oxygenated. With subsidence and marsh flooding may come a decrease in dry upland habitat for diamondback terrapin nesting activites. One of the most dramatic examples of subsidence in the United States has occurred in Galveston Bay, Texas. When aerial photographs were used to compare coastal wetlands in Galveston Bay from the 1950s to 2002 it was found that the total area of salt and brackish marshes has decreased and that there has been some redistribution of marshlands. Due to subsidence, the marshes are being replaced by tidal flats and open water. The major cause of subsidence in the Galveston area may be groundwater pumping around Texas City. With perennial marsh flooding that results from subsidence, there is a serious impact on marsh food webs that can cause foraging problems for diamondback terrapins. In addition, nesting areas may be covered with water year-round or may become more prone to inundation.

Human-engineered structures may impede nesting activities for diamondback terrapins. Various types of seawalls are constructed to "armor" the coastline and thus prevent loss of waterfront homes and property. These structures may be built from wood, stone, concrete, or even sand-filled bags and are called bulkheads or revetments. If these shoreline fortifications are made from randomly strewn rocks, they are sometimes referred to as riprap. While such walls may help protect upland property, they channel wave energy to a narrower band of coastline and thus facilitate the removal of sand and sediments from their base. Over time, sharper embankments are produced and the intertidal beach may be decreased or eliminated. This will decrease potential foraging resources for diamondback terrapins. Although diamondback terrapins are good climbers, the presence of a perpendicular concrete bulkhead is surely a deterrent to a female on a nesting run.

Jetties are stone or concrete structures that are built perpendicular to the shoreline. Groins are smaller versions of these structures, and both interfere with normal littoral drift, that is, the movement of sand and sediments along the shoreline as a result of currents and wave action. Sand tends to be trapped on the upcurrent side of the jetty or groin, while the downdrift side loses sand. The sand-starved beaches and their adjoining uplands may eventually disappear. Since diamondback terrapins usually return to the same nesting locations, year after year, the erosion of a nesting area may force the female terrapin to seek other nesting options, some of which may be less than optimal or represent fringe nesting locations. Nesting may be less successful in marginal habitats that may be more prone to predators, inundation, desiccation, and other forces.

## Aquaculture

The combined threats to diamondback terrapins that have been described are effectively contributing to the sometimes drastic, sometimes subtle demise of the species. There are other potential activities and events that may also threaten diamondback terrapins, but their impact has not been studied. For example, diamondback terrapins share habitat with commercially valuable clams and oysters. After many of the natural shellfish beds became depleted,

*Fig. 5.6. Erosion of dune caused by foot traffic in terrapin nesting area. In nesting areas that are highly vegetated, terrapins prefer to nest on these sandy paths.*

aquaculture developed as an alternative to traditional shellfishing. Aqualculture is often regarded as an environmentally friendly way to utilize and sustain coastal resources. The economic impact in many areas has been extremely positive. Unlike salmon or trout aquaculture, which lead to increases in nitrates in surrounding water, aquaculture is a clean form of fish farming. However, very little is known about the impact of aquaculture on organisms that reside in the mud and marsh substrate, organisms that are important components of the food web and a source of vital fuel energy for migrating birds.

Clams and oysters are raised in concentrated parcels on the tidal flats. Although diamondback terrapins may be oblivious to the activity of the aquaculturists, the trays, netting, PVC pipes, truck traffic, and so on that are part of the industry may impede foraging, interfere with mating aggregations, or provide obstacles to local terrapin movement. Aquaculture gear that is uprooted from the farms during storms may clog creeks and trap terrapins under water.

## Dredging

Recreational and commercial fishermen and boaters are putting pressure on marinas to provide additional mooring and docking facilities and to keep harbors accessible. This sometimes means initiation or increase of dredging activities. Led by C. W. Post biologist Matt Draud, a local Oyster Bay, Long

Island, environmental group called Friends of the Bay has lobbied the town to reschedule dredging plans in Bayville Village at the Creek Beach Marina. Using sonic tracking, Draud's research team has identified the group hibernaculum shared by several hundred diamondback terrapins. Buried beneath the ooze on the floor of the marina, the hibernating terrapins could be easy fodder for the jaws of the giant dredges, scheduled to deepen the marina during the winter to eliminate interference with the boating season. Dredging is a routine procedure at shallow harbors and embayments that strive to provide access for boats. In many areas within their range, very little is known about the winter residence and hibernation areas of diamondback terrapins. Inadvertently, terrapins may be killed as heavy machinery disrupts their winter's sleep.

## Invasive Species

Invasive species have also made their way to coastal marshes and estuaries. As water freshens, *Phragmites* reeds take the place of *Spartina* grasses, flora and fauna shift to freshwater varieties, and diamondback terrapins may lose food sources. Some invasive species, such as *Hemigrapsus* (Asian shore crab), believed to have arrived in ship ballast water and prospering on the Atlantic coast, may outcompete native crab species. The effect of this invasion on the terrapin diet is unknown.

## Recreation

At first glance, it seems lucky for diamondback terrapins that parts of their range fall within wildlife refuges, national parks, state and local conservation districts, and wetland areas that will never be developed. This is a mixed blessing for terrapins, because many of these protected areas are open to the public for hiking, boating, fishing, crabbing, swimming, camping, and other activities (fig. 5.6). Herein lies the conflict. The very people who love nature and appreciate diverse forms of life may inadvertently cause declines in terrapin colonies. Boat strikes, capture in crab pots, snagging on fishing lines and hooks, interference with nesting, attraction of subsidized predators, erosion of nesting beaches, road mortality, and other threats occur just as readily, if not more often, in areas set aside for recreation. Recognizing the interference with terrapin nesting, Jamaica Bay Wildlife Refuge closes certain areas to hikers during nesting season.

In a twenty-year study, conducted from 1974 to 1993, documenting the relationship between human recreation and the decline of a wood turtle (*Clemmys insculpta*), it was seen that the apparently stable turtle population began its precipitous decline when the habitat was opened for recreation. Other parameters of the wilderness area, such as climate, air and water quality, populations of nearby towns, and number of roadways, remained constant. The increase in recreational use of the wilderness area was measured as a function of the number of permits issued each year. Although the cause of the decline was not identified, several mechanisms were hypothesized: removal of turtles by visitors, road kills, handling by visitors, increased predation as a function of attraction of predators by food waste, and disturbance by dogs (Garber and Burger, 1995).

In most cases, we do not know the impact of human recreation on diamondback terrapins. Even the most well-meaning of us may adversely affect the turtles we are trying to save. Clearly we must devote more time and resources to learn how we can eliminate or minimize our deleterious impact on terrapins.

While certain colonies of diamondback terrapins may be holding their own, the broader fate of the species does not look promising. This meek turtle is being threatened from all directions, land and sea. Scientists, naturalists, and conservation-minded citizens along with private, state, and federal organizations are not just sitting back to see what happens to the charismatic terrapin but are taking action to prevent the decimation of our only brackish water turtle.

# Chapter 6

## Learning from the Past;
## Peering into the Future

THERE IS NO EASY PRESCRIPTION for conservation of diamondback terrapins. Although some threats, such as loss of suitable habitat, may affect all subspecies of terrapin in all parts of their range, other impacts are regional. This was apparent when researchers gathered at the Third Workshop on the Ecology, Status and Conservation of Diamondback Terrapins in Jacksonville, Florida, in September 2004. The workshop was organized by Joe Butler from the University of North Florida and George Heinrich of Heinrich Ecological Services. Reports from researchers who attended the conference revealed that terrapins are in trouble. Whether the problem stems from an oil spill, crab traps, road mortality, development, or other factors, a call to action was deemed necessary. The concept of a National Diamondback Terrapin Working Group was formulated by conference organizers Butler and Heinrich along with Willem Roosenburg. At the Third Workshop, the National Diamondback Terrapin Working Group was formed (http://www.dtwg.org) and its mission statement was approved by conference attendees. Members of the group recognized that a national approach to conservation may be limiting due to the wide differences that relate to the different habitats and diverse threats within the terrapin's range. Regional working groups may be more effective. Where should we begin?

### Protective Legislation

Protection of this turtle has taken place in fits and starts. The decline in the diamondback terrapin population as early as the mid 1800s led to concerns about the welfare of the species. Even when it became clear that the numbers of diamondback terrapins were dwindling, the federal government never stepped up to protect these turtles. State governments have taken action to help the population to rebound when local extinction seemed a possibility.

Maryland was among the first states to pass protective legislation. As early as 1878, terrapin harvest was limited to autumn, winter, and early spring. Various states have listed the diamondback terrapin as endangered, threatened, or species of conservation concern. Some states still consider the terrapin a game animal. All states in which the terrapin can be found have adopted some level of protection, from a complete ban on their capture, to limits on the legal hunting/fishing season, number of terrapins that can be harvested, size, and/or methods by which terrapin can be captured. A collection of current regulations with respect to capture and commercial harvest of terrapins has been collated and summarized by Christina Watters (2004) of the Wetlands Institute. The information is presented with some additions and revision in table 6.1.

In some states, diamondback terrapin conservation status and resulting regulations are assessed by wildlife agencies, while in other cases regulation of diamondback terrapins is under the purview of the Fisheries Department. In some instances terrapins are considered a game species, while in others they are included as a species in the commercial fishery. Agencies that regulate the collection of terrapins are often different from agencies that regulate and enforce the use of TEDs or BRDs on crab pots. Most states use a classification system similar to that used by the U.S. Fish and Wildlife Service. According to the information available to the regulatory department, a species is listed as "endangered" if it is in danger of extinction, "threatened" if it is likely to become endangered if current conditions and trends persist, "rare" if the species is potentially at risk because it is found only in a limited geographic area or habitat within its range but not currently threatened or endangered, and as a "species of special concern" if it warrants careful monitoring but does not fall into one of the other categories. As states review their regulations, the trend has been to impose increased restrictions on harvest and collection. When Georgia reviewed its terrapin regulations in April 2003, commercial harvest was completely banned and the terrapin was protected as a nongame coastal resource.

## Demographic Information

Perceptions, based on observations using a variety of techniques, point to declining diamondback terrapin populations in many parts of their range. In order to determine the long-term trends in population size, it is important to have a handle on current population status. In addition to the total number of individuals, a good assessment of population health and stability must include information about age structure, sex ratio, survivorship, population density, and patterns of seasonal dispersal. We must know the distribution of terrapins within localized creeks, bays, lagoons, sounds, estuaries, and marshes. We must know terrapin mating areas, foraging grounds, nesting and hibernation/brumation sites, and nurseries. We should know terrapin food preferences by age class and sex. We need information about seasonal use of habitats. We should also have an idea about age and size classes within colonies. Current information is useful, but demographic trends over time are crucial in determining the conservation status of individual populations. This undertaking is not as easy as it sounds. Although some colonies have been well characterized, large gaps remain in the data.

## Approaches to Population Studies

### MARK-RECAPTURE

In many studies of animal species in their natural habitat, it is often impossible to census all individuals in the population even when the number of individuals is low. To get a handle on the size of diamondback terrapin populations in any particular location, researchers use mark-recapture sampling. Terrapins are captured using various methods, marked using carapace notching, metal tags, and/or PIT tags and are then released into the population. At an appropriate time interval, the population is again sampled and the proportion of recaptured animals during the resampling process can be used to estimate the total population size. In theory, this methodology should provide a robust indication of population size. In order for mark-recapture sampling to provide accurate census data, certain conditions must be met. For example, every animal must have the same probability of being captured in each sampling or resampling event, the number of marked and unmarked animals in the population should not change during the sampling intervals, the marked and unmarked animals must be captured at equal rates, and there should not be a differential probability of capturing animals during the different sampling periods. We can easily see situations in which these conditions will be difficult to meet for terrapin research.

Seasonal, sex-specific, and age-specific use of habitat must be taken into consideration. As research continues, additional marked animals are often

**Table 6.1**  Diamondback Terrapins: A Review of Range-Wide Regulations

| State | Current Regulatory Status and Brief Summary of Regulations |
|---|---|
| Massachusetts | **Threatened:** Illegal to disturb, harass, hunt, fish, trap, or take adults, eggs, or young by any means |
| Rhode Island | **Endangered:** Illegal to buy, sell, or in any way traffic any terrapin or part thereof, either living or dead |
| Connecticut | **State-regulated species:** Open season August 1 to April 30; up to five terrapins 4 to 7 inches allowed; illegal to take eggs |
| New York | **No listing:** Open season August 1 to April 30; terrapins may only be taken by dip nets, hand capture, authorized seine nets, and special traps, labeled with the identity of the owner; license required for take; terrapins 4 to 7 inches allowed; sale allowed May 5 to July 31 or year-round if killed and processed for consumption before May 5; illegal to take eggs |
| New Jersey | **Species of special concern and game species:** Permit required for possession; open season November 1 to March 31; illegal to take eggs; 2 by 6 inch TEDs required on some crab traps, and biodegradable panels required on all traps |
| Delaware | **Species of state concern, SU species (status uncertain), and regulated Game Species:** Open season September 1 to November 15; four terrapins/day limit; illegal to take eggs; legal to raise terrapins in "private ponds"; 1-3/4 by 4-3/4 inch TEDs required on some crab traps; no more than two can be kept as pets |
| Maryland | **S4 species (apparently secure):** Illegal to possess terrapins less than 6 inches (plastron) (but does not apply to those keeping more than three as pets); illegal to take eggs; illegal to harvest or possess May 1 to July 31; 1-3/4 by 4-3/4 inch TEDs required for pots set from private property. Limit of two pots per property |
| Virginia | **S4 species (apparently secure) and listed on Natural Heritage Vertebrate Watchlist:** Illegal to take, possess, import, export, buy, sell, and so on without a permit; cannot be harvested in spring or summer; fishing license required for take; gear restrictions for take |
| North Carolina | **Species of Special Concern and S3 (rare or uncommon):** Take, possession, collection, transportation, purchase, sale of five or more prohibited |
| South Carolina | **No listing:** Closed season April 1 to July 15; terrapins greater than 5 inches (plastron) allowed |

**Table 6.1** Continued

| State | Current Regulatory Status and Brief Summary of Regulations |
|---|---|
| Georgia | **Special concern animal and S3 (rare or uncommon):** No harvest permitted; protected under non-game laws; illegal to keep as a pet regardless of origin or morphology |
| Florida | **No listing:** Gear restrictions for take; illegal to buy, sell, or possess for sale terrapin parts; illegal to possess more than two without a permit; illegal to possess more than fifty eggs without a permit |
| Alabama | **Species of special concern and protected species:** Illegal to take, capture, kill, possess, sell, trade, and so on without a permit. Terrapins must be 6 inches (plastron) |
| Mississippi | **Species of special concern and S2 (imperiled because of rarity):** Commercial traffic illegal without a captive propagation permit (allows capture of up to sixteen animals); legal to possess up to four with a small game hunting and fishing license |
| Louisiana | **Species of special concern and game animal:** Closed season April 15 to June 15; illegal to take by trap; reptile and amphibian collector's or wholesale/retail dealer's license required for sale, barter, or trade; fishing license required for collection; illegal to take eggs |
| Texas | **No listing:** Illegal to possess more than ten without a nongame collection or dealer permit; nongame collection or dealer's permit required for sale or trade; hunting license required for collection |

Note: Some states protect terrapins explicitly while others protect terrapins under more general wildlife regulations. Measurements are given in inches to comply with state regulations.

added to the marked animal roster in a given population. Different capture methods have the potential to produce age and sex biases. For example, gill nets with large holes will tend to retain only females, crab pots may catch only males and juveniles, and nesting surveys will census only mature females. Numbers of smaller juveniles are typically underestimated because of their cryptic behavior and utilization of upland areas of the marsh. Furthermore, any immigration into the colony or emigration from the colony will affect the population analysis.

Various statistical models have been developed and employed to estimate diamondback terrapin population size and structure from mark-recapture data while taking the problems with sampling into consideration. As long as researchers consider the limitations of their models and interpret population data accordingly, useful and relevant information can be obtained for conservation and management purposes.

Using mark-recapture data from 1990 to 2001 for approximately 188 nesting females in the only known diamondback terrapin colony in Rhode Island, Mitro (2003) was able to devise a growth model and calculate a population growth rate of 1.034, indicating a relatively stable population. This study was also able to point to individual female survival rather than recruitment of breeding females as the probable cause of the stability of the population. Even though a stable population might signal that this lone cluster of Rhode Island terrapins is not in immediate jeopardy, the low level of recruitment suggests that this population will decline as the females continue to age and eventually die.

Working with Peter Auger, Hart used sixteen years of mark-recapture data to analyze population parameters for diamondback terrapins in Barnstable, Massachusetts (Hart, 1999; Hart et al., 2000). This population is considered to be relatively stable and is not threatened by commercial harvest or mortality in crab pots. Using various sampling methods, 452 juveniles and adults were captured and 449 of these individuals were recaptured in a 500-square-meter (approximately 600 square yards) site. Using various assumptions about the stability of the population, age class-specific survival rates, and literature values for other turtle species, it was estimated that adult terrapins in this location have a survival rate of 0.83, juveniles have a survival rate of 0.60, and hatchlings have a survival rate of 0.23. To determine how crab pot mortality may affect the population over time, this population scenario was applied to a hypothetical model in which a certain percentage of the population, particularly the juvenile cohort, dies each year as a result of crab pot mortality. The model predicted that there can be a drastic impact on the population when juvenile survivorship is impacted by crab traps. The rate at which this long-lived population would dramatically decline could be predicted as a function of the extent of crabbing operations.

A long-term population study of the diamondback terrapin has been conducted by Whit Gibbons, Jeff Lovich, Tony Tucker, and other colleagues at the Savannah River Ecology Laboratory, University of Georgia at Kiawah Island, South Carolina. The study site is near a 3,200-hectare (about 8,000 acres) barrier island in a mild climate. Half of the island is a salt marsh that is subject to 2-meter (2.2 yards) semidiurnal tidal variation. *M.t. centrata* subpopulations have been studied at this site for twenty years using mark-recapture techniques. Adult terrapins were captured in four tidal creeks, approximately 1 to 2 meters (1.1 to 2.2 yards) apart. Kiawah Island juveniles utilize a different habitat. Although various capture techniques were used, including baited hoop traps, dip nets, and trawling, the most effective methods for this site employed trammel nets and seine nets with mesh sizes that were designed to capture adults and larger juveniles (Tucker et al., 2001; Gibbons et al., 2001). Over the course of the study, there were more than 1,200 original captures and more than 1,100 recaptures. This creek system, which includes Oyster, Terrapin, Fiddler, and Sandy creeks, is considered to harbor a terrapin metapopulation, which may be typical for other diamondback terrapin colonies as well. A metapopulation is a single population inhabiting a large area but fragmented into several smaller subpopulations; the individuals are dispersed in patches. There may be a potential, depending on the ecological circumstances, for specific patches to decline or even become extirpated. However, there may also be the potential for repopulation by individuals from nearby subpopulations.

The mean survival rate for the entire metapopulation was 0.835 for males and 0.84 for females, with rates somewhat variable among the creeks. To assess metapopulation dynamics, Tucker et al. (2001) examined the potential movement of terrapins between creeks. A creek-by-creek analysis of the population indicated that there was strong site fidelity, with overall low rates of transition of terrapins from one creek to another. Over a seventeen-year time period, a significant population decrease was observed in Terrapin Creek. This decline was attributed to incidental capture of terrapins in crab pots and subsequent drowning. This mark-recapture study resulted in a calculation of the net probablilty of males moving from creek to creek of 0.0 to 0.044 and for females 0.0 to 0.215. These movement probabilities suggest that in this system, females are more likely to move between creeks than males. With overall low rates of movement, the researchers speculated that an extirpated cluster within the metapopulation was not likely to be repopulated from terrapins that inhabit nearby creeks. This population modeling study also suggested that there was a mean life span for female terrapins of 5.7 years, which is below the age at which females in this population reach maturity. In speculating about the cause of the low female survivorship, it was noted that females spend more time in deeper water, an area of high boat traffic. More females than males have been found with propeller injuries (Gibbons et al., 2001). Predation by raccoons was also thought to contribute to low female survivorship. Adults in this population are already under severe pressure due to crabbing, especially after the construction of a dock at Inlet Cove, across from Terrapin Creek, in 1983. Initially, a male bias was reported for this population. However, by 2004, there had been a shift to a more female-biased ratio. Clearly, crab pot mortality is taking a severe toll on males and smaller terrapins. Individuals from younger age classes are more and more difficult to find (Gibbons et al., 2004).

By 1993, the Terrapin Creek subpopulation was considered extirpated. Terrapins have not returned. Due to the high degree of site fidelity, this local extirpation has created a void that may take a generation or more of recruitment to fill, even if the continuing threats are eliminated. It therefore seems unlikely that this colony can be repopulated by terrapins from neighboring creeks, especially if the cause of the decline is not eliminated. As a result of the dramatic decline in diamondback terrapins and the unlikelihood that terrapins will return to the area in the near term, perhaps Terrapin Creek should be renamed.

### THE POWER OF GENETIC STUDIES

Modern genetic studies use a variety of molecular tools and techniques for phylogenetic and ecological assessments. Phylogenetics refers to the description of evolutionary or ancestral descendent relationships among organisms. Genetic information is widely accepted as a valuable tool in constructing such evolutionary relationships. A phylogenetic approach to reptile evolution using DNA sequence data provided the key information that placed turtles and crocodiles closer together on the reptile evolutionary tree (chap. 1).

Different types of molecular markers are available or can be developed for phylogenetic and population genetic analysis. These markers are regions of DNA that display variation within and among species. Some markers have been found by determining the nucleotide sequence of a particular gene or gene region and then comparing the sequence in the same region from species to species or even organism to organism. Other markers, known as restriction fragment length polymorphisms (RFLPs) are produced as fingerprints that are generated after severing DNA with specific enzymes known as restriction enzymes. Individuals with the same DNA sequences will produce the same fingerprints; differences in DNA sequence are detected by alterations in the fingerprint pattern. An example of a gene region that proves to be useful in this type of analysis is the control region within the D-loop of mitochondrial DNA. This particular section of the mitochondrial genome has been shown to be evolving at a high rate, thus exhibiting a great deal of variation. Mitochondrial DNA (mtDNA) takes the form of a small, circular chromosome and is contained within organelles known as mitochondria, where the energy-generating actions of cells occur. A unique characteristic of mitochondrial DNA is that it is maternally inherited. Each turtle (or human, for that matter) gets most of its mitochondria from the egg produced by its mother. Sperm do not contribute significant mitochondria to the fertilized egg, so the father's genes are not represented in mitochondrial DNA. Mitochondrial DNA markers have been very helpful in tracing the global migration patterns of sea turtles and in supporting the natal homing hypothesis. Using mitochondrial D-loop markers, biologists were able to show natal homing in sea turtles; females return to the same beaches from which they were hatched to lay their own eggs. Such information has highly significant management and conservation applications.

Mitochondrial markers have been employed in phylogenetic analysis of *Malaclemys terrapin*. A high level of genetic exchange within the species would be expected due to its geographically continuous habitat range from Massachusetts to the Mexican border. As described in chapter 1, morphological differentiation has indicated seven distinct subspecies. In searching for molecular differences within 16,800 base pairs of terrapin mitochondrial DNA, Lamb and Avise (1992) used RFLP analysis and focused on the D-loop. The data show a very low level of DNA variation compared to other vertebrates that have been examined using similar methodology. Only one marker was found to differentiate diamondback terrapins. The marker separated terrapins into two groupings with the dividing line drawn in central Florida, near the Kennedy Space Center. The analysis separates subspecies *terrapin* and *centrata* from the other five subspecies, *tequesta*, *rhizophorarum*, *macrospilota*, *pileata* and *littoralis*. Mitochondrial genetic analysis followed the differentiation pattern displayed by a single morphological characteristic: the tuberculate keel on the midline of the carapace, which is more common in southern subspecies. Genetic analysis did not distinguish populations that differ in other morphological characteristics or by fine-scale geographic distribution patterns.

Mitochondrial DNA, in particular the control region, which is rapidly evolving, and the cytochrome *b* gene, which is evolving at a moderate rate, were sequenced in order to provide evidence for the evolutionary relationship between *Malaclemys* and *Graptemys* (map turtles) (Lamb and Osentoski, 1997). In the same study, comparisons were made among mitochondrial DNA sequences from five of the seven subspecies of *M. terrapin*. Similar to results of the molecular study by Lamb and Avise (1992) and the morphological examination by Carr (1946), DNA sequence comparisons produced a clear differentiation between the northern subspecies (*terrapin* and *centrata*) and the three southern subspecies that were examined (*rhizophorarum, pileata,* and *macrospilota*).

Studies in various turtle species indicate that evolution of mitochondrial DNA is particularly slow compared to other species and that local populations do not appear to be genetically unique when mitochondrial markers are used (Avise et al., 1992). The same phenomenon was described when mitochondrial genes were analyzed to examine the effect of habitat fragmentation on genetic variation in the bog turtle, *Clemmys muhlenbergii*, throughout its range from southern Massachusetts to northern Georgia. Although morphological criteria differentiate a northern and southern subspecies, a mitochondrial marker, 16s ribosomal mtDNA, failed to detect any variation in the dispersed colonies or between the subspecies (Amato et al., 1997). There may be various reasons for the slow rate of evolution of the mitochondrial genome in turtles. The long lives of turtles result in fewer generations in a given time period. Furthermore, some turtles may have dispersed relatively recently, since the last glacial retreat approximately 15,000 years ago.

In many cases, it is also important to obtain markers that are evolving at higher rates than mitochondrial genes and represent the genetic contribution of both males and females in the population. For this purpose, biologists search for markers within nuclear genes contained on chromosomes that have genetic information from both parents, contributed in equal amounts. If a nuclear sequence of DNA is invariable, as it is for most genes that specify the production of our proteins, it will be the same in all organisms within the species and will not be a useful probe to assess variation, taxonomy or ecology.

The value of variable markers extends beyond population genetic analysis. They can also be used to identify the geographic origin of individual turtles and assign each turtle to their natal population. This feature of genetic analysis may be helpful in enforcing harvest and collection regulations. To find the type of marker that displays variation, biologists look to regions of the genome that lie outside of coding regions. The function of these regions is generally not known, but they appear to contain quite a bit of variation. There are several types of nuclear markers, but the category that has been used in diamondback terrapin studies is called microsatellite DNA. This category of marker contains short stretches of nucleotides, generally one to five, that are repeated many times. The number of tandem repeats will determine the length of the sequence. There may be many forms of a particular microsatellite marker within the general population, but each individual can only have two forms or alleles, one from its mother and one from its father.

Microsatellite markers can be used to assess genetic variability within and between diamondback terrapin populations. Noting the lack of sensitivity of D-loop and other mitochondrial markers to distinguish terrapin populations, Miller (2001) used microsatellite markers to test whether Hurricane Georges caused dispersal of terrapins from *M. t. rhizophorarum* colonies in Florida. By examining allele frequencies in the populations inhabiting specific locations, before and after the hurricane, she concluded that the storm did not cause dispersal or redistribution of terrapins.

An attempt to understand the impact of the geographic distribution of a species on its phylogeny is called phylogeography. Molecular techniques can be used to create population models that will link geographic distribution patterns, genetic characteristics, and demographic features. This approach can measure the extent of gene flow or genetic exchange between individuals in the population or among the subpopulations. Such information can inform researchers and wildlife managers whether certain geographical groups of terrapins form one large population/subpopulation or whether management efforts must focus on preserving separate, unique colonies.

In separate studies, researchers Kristen Hart and Susanne Hauswaldt used a genetic approach to complement their studies of diamondback terrapin ecology. While a student at Duke University, in the program in ecology, Hart used microsatellite markers developed in the USGS laboratory of Tim King (King and Julian, 2004) to define boundaries of diamondback terrapin populations (Hart, 2004). Although the markers were initially developed for bog turtles, they identify polymorphisms or genetic variation across a wide range of turtle species. Hart's analysis identifies six regional populations or groups of diamondback terrapins that break down as follows:

1. Northeast: Rhode Island and Massachusetts.
2. Coastal mid-Atlantic: New York, New Jersey, and Delaware.
3. Chesapeake.
4. Coastal Carolinas.
5. South Florida: south Atlantic coast, Florida Bay, and Florida Keys.
6. Gulf of Mexico.

Within each grouping, several fine-scale distinctions can be made. The genetic picture suggests that females remain in local areas, a situation known as philopatry, while males are responsible for dispersing genes among sites within the metapopulations. In other words, the males are moving about much more than the females.

This type of information has the potential to be employed in conservation and management. For example, the regional groupings may be so genetically similar that it may be possible to design repatriation and relocation strategies within a group to increase the number of terrapins in colonies that are experiencing decline.

Using microsatellite markers she developed (Hauswaldt and Glenn, 2003, 2005) as well as D-loop sequences in mitochondrial DNA, Hauswaldt's studies at the University of South Carolina were focused on a fine-scale population genetic analysis of terrapins within specific estuaries. She hypothesized that the high site fidelity displayed by terrapins would be reflected in the population genetic structure. Using Charleston Harbor as a study site, Hauswaldt was not able to differentiate terrapin clusters on the basis of their genetic profiles. She found no significant difference between sites within a river, between males and females, between terrapins during different seasons, and even among different rivers feeding into the Harbor. She concluded that the high site fidelity of terrapins is not reflected in their population genetic structure. Her study also led to the conclusion that male terrapins are responsible for gene dispersal within the Charleston Harbor cluster.

Hauswaldt's genetic analyses also pointed to a curious finding: East Coast terrapins are genetically more similar to Texas terrapins than to Florida subspecies. She speculates that this is due to the intentional mixing of the subspecies in the early 1900s. For example, *M. t. littoralis* was imported to Maryland and the Carolinas and used for hybridization with the supposedly tastier Chesapeakes during the heyday of captive breeding programs. The release of Texas turtles or farm-raised hybrids into Atlantic waters for restocking purposes has led to the spread of Texas terrapin genes on the East Coast. Hauswaldt also used her microsatellite markers to demonstrate multiple paternity in diamondback terrapin clutches (discussed in chapter 3). Using hatchlings and females from Matt Draud's study site on Long Island, Hauswaldt analyzed 437 hatchlings from twenty-six females, including seven double clutches. She found evidence for multiple paternity in six out of thirty-three clutches. In some cases, only a few hatchlings were fathered by a different male. An interesting finding was that no females were shared among the males, suggesting a large breeding population (Hauswaldt, 2004). Thus it appears that multiple paternity occurs in diamondback terrapins but is relatively rare compared to other turtle species.

As powerful as genetic studies may be, there are some issues and questions that this approach alone may fail to answer and other questions that can be raised. Both Hart and Hauswaldt found very low levels of genetic variation within terrapin clusters and also within relatively large geographic areas. The low levels of genetic variation within clusters support the many ecological observations regarding site fidelity. However, the low range-wide variation seems to contradict ecological field studies that uniformly indicate that terrapins, in general, are nonmigratory and stick close to home. Such behavior is expected to produce inbreeding and thus genetically unique populations.

These contradictory findings bears some explanation. Inbreeding is generally regarded as an unfavorable process for a population. With inbreeding, there is always a chance that deleterious genes will be found in greater frequency and that this can eventually lead to the demise of a population. Although there might be a high level of inbreeding in terrapin populations, microsatellite analysis reveals that there is some gene dispersal occurring due to movement of males; local populations are not quite as genetically unique as predicted from the ecological analysis. Furthermore, there are many cases of species that are highly inbred but whose populations have rebounded, at least in numbers, from various environmental or anthropogenic insults. We can also look at genetic diversity in another context. Even if terrapins exhibited a high degree of genetic diversity, the population would still be threatened if the numbers of individuals were so low that they couldn't find mates or if sexes were disproportionate in an unfavorable ratio. Therefore, the jury is still out regarding the potential impact of the relatively low genetic variation within local populations of diamondback terrapins as well as in terrapins throughout their range. Ecological and behavioral field studies are also required before we can fit together all the pieces of the population puzzle.

## Habitat Restoration

A significant amount of diamondback terrapin nesting habitat has been irretrievably lost. Homes, seawalls, resorts, and highways take the place of sand dunes and beach strands. It is logical to assume that diamondback terrapins would utilize new or restored nesting areas if they are close to original nesting areas and placed within the range of the local population. This has certainly been the case at the Horsehead Wetlands Center in Maryland where a small nesting beach was created. Using truckloads of sand, Marguerite Whilden worked with coastal engineers and landscape architects to create a nesting beach (fig. 6.1). The beach is situated shoreward of an artificial oyster reef that was constructed by the Maryland DNR. The reef shelters the beach and will mitigate against erosion. Females nested on the beach in the first year it was created.

This pilot program for creation of nesting habitat will provide valuable information that can be applied to other conservation efforts. Nesting beach creation or renourishment efforts not only benefit diamondback terrapins but also shorebirds and other organisms. However, these attempts must be made with a thorough understanding of local coastal geology and sediment transport. What good is it to restore or renourish a beach, only to have it wash away in a year or two? Furthermore, nesting beach restoration by itself will not be a useful conservation approach if other threats to diamondback terrapin populations are not addressed. At sites such as the Brigantine Wildlife Refuge in New Jersey, now part of the Edwin B. Forsythe National Wildlife Refuge, adequate nesting habitat alone was not sufficient to maintain diamondback terrapin populations (Burger and Garber, 1995).

In some cases, the aqueous or marsh habitat of the diamondback terrapin

*Fig. 6.1. Habitat restoration project at the Horsehead Wetlands Center, Chesapeake Bay. A nesting beach was created near an artificial oyster reef.*

has been altered. Restoration of these areas can also be very important in conservation efforts. The proposed change in tidal flow of the Herring River estuary in Wellfleet, Massachusetts, is an example of an attempt to restore a brackish water and marsh system. For mosquito control purposes, a dike was built across the mouth of the Herring River in the early 1900s. Upstream waters are fresher than they were before the dike was built; downstream, the water has become very saline. Opening the dike gates will partially restore the historic flow of the Herring River and is expected to slow down ongoing subsidence of the marsh. The project is not specifically targeted for diamondback terrapins. Instead, it represents an ecosystem approach to conservation and is expected to affect many species of plant and animal. As marsh conditions are restored, the project will undoubtedly have a positive impact on the terrapins that utilize this estuary.

## Crab Pot Mortality

In many cases of diamondback terrapin conservation, long-term demographic data are important but the writing is on the wall; there may be an urgency with respect to action. Crab and eel pots are killing many terrapins. Commercial crab potting is prohibited in the Patuxent River, but local homeowners can place pots in shallow water adjacent to their property. As a result of a habitat utilization study in the Patuxent River, Roosenburg et al. (1999) have suggested that these shallow waters are precisely the areas used by younger, smaller terrapins, with resulting crab pot mortality. It has been suggested that crab pots not be allowed in shallow water. At the very least, these shallow-water pots should be fitted with BRDs or have a structure that maintains permanent air space (Roosenburg et al., 1999).

How much data are needed to demonstrate that terrapin populations are declining? How many terrapins must die in crab pots before regulations are adopted? With data from field studies and an uphill battle with state regulators, it was possible for Roger Wood to champion the use of terrapin excluder devices in New Jersey (fig. 5.5). A 1998 regulation mandates the use of TEDs (BRDs) on commercial crab pots that fish in waters less than 150 feet wide at low tide or in man-made lagoons. Wood would like to see the regulations extended to include recreational crab traps as well. Regulations have also been enacted in Delaware and Maryland. In Maryland, recreational crab pots must be fitted with BRDs, but commercial crabbers are exempt from the regulations (table 5.1).

In 2003, the Environmental Committee of Kiawah Island, South Carolina, offered to outfit crab pots, free of charge, with BRDs to protect terrapins. The program is completely voluntary and designed as a conservation measure for recreational crabbers.

Several state and local environmental groups have taken initiatives in removing ghost crab pots. Crab pots are private property and cannot be indiscriminately removed from the water, even after they have drifted from their original moorings. Since 2002, the Texas Parks and Wildlife Department has conducted an abandoned crab trap removal program during a ten-day moratorium on crabbing each February. Any pot that remains in the water during the ten-day period is assumed abandoned. Aerial surveys are used to determine locations of abandoned gear. Over a three-year period more than 15,000 pots were removed with the assistance of volunteers (Morris, 2004). In 2003, the Gulf States Marine Fisheries Commission developed a regional program to remove traps from Alabama, Mississippi, Louisiana, and Texas. Thus far, 30,000 derelict traps have been removed (Perry, 2004). Ghost crab pot cleanups are being organized and implemented in states along the mid-Atlantic region, where thousands of them, sitting in 1.5 to 3 meters (5 to 10 feet) of water, can be seen from the air.

Scientific studies indicate TEDs/BRDs will not adversely affect crabbing operations. In some cases, more crabs are caught in traps with exclusion devices than without them. The equipment is relatively inexpensive and will prevent diamondback terrapin mortality and will lessen by-catch of other species as well. These devices will not only be important on crab pots that are actively being used but become even more significant on ghost traps. Some crab traps have even been fitted with biodegradable panels to allow free egress of creatures that might otherwise become entrapped in ghost pots.

## Nest Protection

At several diamondback terrapin research sites, eggs and nests are being protected with metal or plastic devices of various designs. Sometimes these nest protectors are as simple as a square of wire mesh, tacked down around the nest. Some contraptions are fabricated from hardware cloth, wire, or plastic mesh (fig. 6.2, top). Other designs are large exclosures in which the eggs from multiple nests can be relocated and allowed to develop in a protected environment (fig. 6.2, bottom). These devices are employed to prevent depredation of

***Fig. 6.2.*** *Design of nest protectors. Some predator excluders protect individual natural nests (top), while others provide a protected area for nest relocation (bottom).*

nests by raccoons and other nest predators. They are sunk into the substrate around the nest to hopefully prevent predators from digging under the structures and getting at the eggs.

In order to use a nest protector or predator exclosure it is important to be able to locate freshly laid eggs. This is possible using field marks (fig. 3.1) or by observing nesting activity. Sometimes it is important to confirm the presence of eggs before making the effort to protect the nest. We have known diamondback terrapin females to make false nests; they go through the entire nesting ritual but do not deposit any eggs in the nest. Once the nest is protected, it must be monitored throughout the spring and summer to make sure that a predator does not dig under it. I have performed daily checks on protectors placed over nests that are on the margins of roadways or parking lots, replacing them as necessary during the summer as they get squashed by cars or removed by vandals. More important, nest protectors must be monitored during hatching season to assure that hatchlings do not become trapped and that they remain able to make their escape into the marsh.

*Fig. 6.3. Brochures inform citizens about terrapin conservation projects and provide contact information for reporting terrapin sightings.*

Other research and conservation groups have developed their own designs for nest protectors and have had success with their employment. Some of us became concerned that these cages and bins would alter nest temperatures and thus skew the ratios of male to female hatchlings. After spending a few years burying temperature loggers in protected nests and the adjacent substrate, we are convinced that the somewhat manipulative effort of protecting nests will not change the sex ratio of hatchlings in the nests. We don't see a significant difference in diel or seasonal nest temperatures under exclosures as compared to adjacent soil (fig. 3.4) (Brennessel and Lewis, personal observations).

On Cape Cod, we have had tremendous success with wire cage nest protectors for a small percentage of viable nests that we are able to find. Only a handful of exclosures have been overturned or excavated by raccoons. Some protected nests succumb to maggot infestation or root predation, but most produce viable hatchlings in late summer and fall. Other conservation groups have had similar success with nest protectors. When used to protect natural nests, the efforts to protect developing eggs are not as artificial or manipulative as taking eggs from nests and incubating them in the laboratory.

## Local Conservation Efforts

Individuals, environmental groups, schools, and state agencies have all been involved in local and regional activities on behalf of diamondback terrapin conservation. There are many similar initiatives, including terrapin hotlines, "adopt a turtle" or "adopt a hatchling" programs, recruitment of volunteers to monitor nesting, and academic research to fill in those demographic and life history gaps. Many groups have produced educational brochures to inform the public about the problems that terrapins are facing (fig. 6.3). To highlight these efforts, I describe a few of the more visible and enduring programs and some of the individuals involved.

### LOCAL ACTION: NEW JERSEY

While New Jersey male and juvenile terrapins are drowning in crab traps, females, and the eggs they carry, are being flattened by cars. Under the direction of Roger Wood and Roz Herlands, the Wetlands Institute has taken the lead in raising public awareness regarding terrapin road mortality. The emphasis has been on community education and involvement. Campaigns such as "Drive Eggstra Carefully" alert motorists to terrapins on the roads and causeways as female turtles seek nesting locations. Interns and volunteers scour the roads to look for crossing terrapins and place them out of harm's way. Exhibits and displays, the "Teaching about Terrapins" program, events at local schools, and community functions also alert community members and vacationers to terrapins on the roads. The dramatic "Road Mortality" display, which gives the seasonal tally of road-killed terrapins, averaging about 500 per season, is updated daily. In attempts to restore the numbers of female terrapins that are killed by autos, eggs are removed from the dead animals and incubated in the laboratory at temperatures that will produce female hatchlings. These females are then returned to the marsh.

Roger Wood, who spearheads these efforts, has been recognized as a terrapin hero. At the Third Workshop on the Ecology, Status, and Conservation of Diamondback Terrapins in 2004, Wood was presented with an award to honor his many years of diamondback terrapin research, education, and conservation.

## LOCAL ACTION: MARYLAND

Marguerite Whilden is a lightning rod in Chesapeake Bay terrapin circles, igniting controversy with her approaches to diamondback terrapin conservation. I caught up with Whilden on Maryland Day in April 2004, a day of celebration held under tents and on the beautiful campus greens at the University of Maryland in College Park. This day of learning and fun is attended each spring by tens of thousands of University of Maryland students, friends, parents, alumni, prospective students, and residents of the Baltimore and Washington areas. It was entirely appropriate that the University of Maryland mascot be represented. The official mascot since 1932 is a very famous terrapin, named "Testudo." Athletes from the University of Maryland are referred to as "terrapins," and "Fear the Turtle" has become a school rallying cry.

The cartoon-like Testudo

*Fig. 6.4. Testudo balloon at the University of Maryland, College Park.*

was there in full force: as a giant air filled balloon (fig. 6.4), and displayed proudly on hats, tee-shirts, and sweatshirts. A portion of the proceeds from the sale of this "Fear the Turtle" merchandise is funneled into terrapin conservation and habitat restoration projects. An elegant, lifelike, bronze sculpture of Testudo is displayed at the entrance to the McKeldin library, a gift from the class of 1933 (fig. 6.5). Folks were lining up to rub the glistening terrapin's head, a gesture assured to bring good luck, and to have their photo taken with the metallic reptile.

Live terrapins of all sizes were brought to campus by Whilden and her UM terrapin intern for display in the Big Tent, where a terrapin roaring contest was underway. In a new series of television commercials, the UM terrapin roars like a lion, a type of publicity campaign to promote school spirit and pride. One of Whilden's terrapins was hamming it up for the cameras, posing with arched neck as she was passed from admirer to admirer. This female had been injured and was being rehabilitated, but she clearly preferred life in the limelight to life in the wild. Live terrapins could also be found at the plaza, where alumni events were occurring. Here, the terrapin display had to compete with a contest to select "Maryland Idol," a takeoff on a popular television show, *American Idol.*

Whilden worked for 30 years with various environmental agencies, including the Maryland Department of Natural Resources, and was instrumental in developing and implementing a novel, multifaceted terrapin conservation and education program. The DNR initiatives, primarily directed toward commercial fishermen, were expanded to include a larger audience. Instead of simply focusing on the use of BRDs and licensing of terrapin fishermen, Whilden extended outreach efforts to target local citizens. She galvanized a large group of individuals to become members of her "Terp Team": watermen,

*Fig. 6.5. Bronze Testudo outside McKeldin Library, gift to the University of Maryland from the Class of '33.*

veterinarians, landscape architects, property owners, boat owners, welding companies (to make the BRDs), donators of reptile incubators, tanks, and other equipment, and many local elementary schools.

Terrapin Station was Whilden's brainchild. It was established in 1998. One of its programs, Turtle Tots, brought terrapins to the classroom. Eggs from "compromised" nests were hatched and then headstarted by young children. Raising terrapins was a jumping off point for teachers to incorporate lessons about conservation and to educate the next generation of environmental stewards. One initiative was implemented in the environmental education program for third graders in Calvert County public schools. As the school children assumed the responsibility of temporary terrapin guardians, they became interested in the turtles and excited about preserving the Chesapeake Bay habitat in which terrapins live. In the spring, on Terrapin Day, the headstarted terrapins were released at the locations where their eggs were found. At an event dubbed the "Chesa-Peake-ness," young diamondback terrapins raced back into their natal marshes. The headstarting program was eliminated when a few individuals expressed fear that the terrapins might transmit *Salmonella* to the children. Some scientists were worried that the classroom-reared terrapins might harbor pathogens that could not be detected by the screening program that was being used to assess their health and, as a result, would introduce these pathogens into the wild populations.

During her tenure with the Maryland DNR, Whilden, the "Turtle Lady," developed partnerships with many local organizations, including the Severn River Association, Sherwood Forest Naturalist Program, Wildfowl Trust of North America, Assateague Coastal Trust, Whitehall Bay Institute, many private citizens, and the U.S. Naval Academy in Annapolis, where a terrapin nesting sanctuary was created. She encouraged shorefront property owners to post "Terrapin Sanctuary" signs on their bayfront beaches and worked with them to search for ecofriendly alternatives to hardening their shorelines.

The Living Shorelines project, developed by the Maryland DNR under the leadership of Kevin Smith, is targeted to waterfront property owners and helps them to devise a strategy for protecting and preserving shallow water habitat adjacent to their homes. Some owners have even taken the expensive plunge to purchase tons of sand and "restore" their shorelines. Landowners are advised and assisted with creating critical-area buffers of 100 feet. These buffers consist of naturally vegetated areas landward from the mean high water mark of tidal wetlands and tributary streams. Such buffers function to filter sediments, nutrients, and toxins that might otherwise be dumped into the bay as a result of runoff, and the buffers provide critical habitat for terrestrial and aquatic species. The program thus promotes active shoreline stabilization with the use of non-structural erosion control methods. Marsh grasses such as *Spartina* and other vegetation and natural landscaping materials take the place of seawalls, bulkheads, and riprap. Whilden has promoted the Living Shoreline project as a terrapin conservation mechanism. "Garden, Don't Harden" is her slogan for this method of shoreline and beach strand preservation and restoration. Promotion of landowner stewardship is sometimes a hard sell. These programs can be quite costly and force landowners to think beyond the needs of preserving their own property.

For the uninitiated citizen, Whilden describes the diamondback terrapin as the buffalo of Chesapeake Bay, an exploited creature whose numbers have declined as a result of human activities; terrapins, like extant buffalos, are dependent on us for their protection and survival. Whilden worked with the University of Maryland on the "Fear the Turtle" campaign, and she has also helped with the establishment of diamondback terrapin nesting sanctuaries. The terrapin sanctuary in Crisfield is near the site of one of the original terrapin farms.

If anyone can develop a catch phrase to attract individuals and groups to the topic of diamondback terrapin conservation, it is Whilden. With the Maryland DNR, The Diamondback Terrapin Task Force, and other partners, she has initiated campaign after campaign and sponsored event after event to attract folks to various conservation initiatives. She feels that her terrapin campaigns not only will benefit the Maryland State Reptile, but also have the potential to rally conservation-minded citizens to protect and restore Chesapeake Bay. Whilden was also a founding member of the Terrapin Research Consortium, a research advisory group "convened independent of political, academic and government organizations to develop sound research, management and educational standards and share information freely, fully and accurately in the interest of the Diamondback Terrapin resource throughout its range" (from the Terrapin Institute web page: http://www.terrapoininstitute. org/consortium.htm).

When budget cuts eliminated Whilden's DNR position as program manager for terrapin conservation, she remained undaunted. Her efforts have not lost momentum. She founded a private, not-for-profit terrapin conservation organization, the Terrapin Institute, in 2003 to continue projects that were initiated when she worked for the DNR and to move in additional conservation directions. She obtained funding from the energy company PEPCO, the

University of Maryland "Fear the Turtle" fund, and many private individuals. "Marsh Madness" is one of her campaigns aimed at UM students and alumni/ae to protect their school mascot. For a $5.00 contribution, a market terrapin is purchased, tagged, and released in a person's name. The donor gets a certificate of sponsorship and the terrapin's tag number.

Aside from her controversial purchasing and releasing of market terrapins from Chesapeake watermen and seafood wholesalers, described in chapter 5, she remains involved in numerous research, education and conservation initiatives to preserve Maryland's State Reptile. She is backing a proposed five-year moratorium on commercial harvest so that a large-scale study can be conducted to determine the Chesapeake Bay terrapin population size and distribution. She remains concerned about the increasing pressure to harvest terrapins for Asian markets. Although commercial terrapin fishing is not a large-scale operation, watermen continue to catch terrapins as unreported by-catch, and they may have less incentive for returning these animals to the bay if eager buyers are in the wings. With increasing demand for turtles, more watermen may be lured to the terrapin fishery. Without a good handle on current harvest information, it will be difficult to follow any changes in the commercial trends. Whilden pitches conservation to watermen and urges them to report any turtles they take that have her tags attached. As added incentive to report, she guarantees their privacy.

A USGS team from Patuxent Wildlife Research Center has launched a significant undertaking: a study to assess the current status of the bay's terrapin population. Using fyke nets, baited traps, and the help of watermen to locate terrapins in winter hibernacula, the team has captured and tagged hundred of terrapins. Although Chesapeake Bay is a large area to cover, the team is hopeful it will be able to produce an accurate assessment of the status and distribution of the "Buffalo of the Bay," which will inform future conservation efforts.

### LOCAL ACTION: MASSACHUSETTS

Don Lewis of Wellfleet, Massachusetts, can be described as the Cape Cod motivational speaker for terrapin conservation. Lewis has received a considerable amount of local press for his work with Cape Cod terrapins and has raised the level of awareness for the terrapin's plight. Always emphasizing the positive, Lewis gets people excited about diamondbacks and preservation of their habitat. His road show to schools, herpetological societies, Wellfleet visitors, and other groups connects people to terrapins in ways that make folks care about these turtles. Lewis and Sue Nourse, a marine science teacher at Tabor Academy in Marion, Massachusetts, have created and implemented a teaching and research module using "Turtles in the Classroom" as a theme. Their model, "Saving Turtles from Estuaries to Deep Blue Seas," partners students with scientists. In this kindergarten to Grade 12 model, students become research investigators; they collect field data, take measurements, and participate in various research projects. Older students make presentations to elementary schools and explain ecological concerns to local community groups. Tabor is situated on the waterfront, and Tabor students are lucky enough to have diamondback terrapins right in their own backyard. The collective efforts of Nourse and Lewis confirmed the presence of diamondback terrapins in Sippican Harbor and Aucoot Cove, and have identified diamondback terrapin mating aggregations and nesting areas. Using the curricular ideas they developed and piloted for Tabor students, Nourse and Lewis have adapted their curriculum as a way for other schools to introduce science concepts in a hands-on, interactive manner. When Nourse explained the model to me, she described the "delight for an educator" when students become excited about science. Referring to the terrapins, she commented. "The best part of the whole turtle program is watching their magic transform the classroom."

There are some who have taken environmental action in their own hands, but their motivation is not conservation and the impact of their actions may not be beneficial to the Massachusetts terrapin populations. It is an annual rite of New York City Buddhists to practice a "releasing life" ceremony. Animals that are destined to be killed for human consumption are purchased and freed. In June 2004, Manhattan Buddhists purchased about 300 diamondback terrapins from markets in Chinatown and transported them to the village of Padenarum on Buzzards Bay. After suitable "releasing life" prayers were chanted, the terrapins were set loose into Massachusetts waters. No one knows the origin of the terrapins; most likely they were from the mid-Atlantic region. No one knows if the released turtles will settle into their new habitat and breed with the local population. No one knows how long the Buddhists have been releasing Chinatown terrapins into Buzzards Bay. No one knows if these "foreign" turtles will confound the fine-scale terrapin genetic studies that are being planned for the southwest coast of Massachusetts.

## LOCAL ACTION: FLORIDA

George Heinrich believes that biologists have an obligation to conduct or participate in conservation education programs. Educators are conduits for dissemination of information to the public, policymakers, and the next generation of environmental advocates. With this in mind, he has used Florida's turtles, including diamondback terrapins, as a centerpiece for environmental education workshops for educators. Educators participate in classroom activities that are supplemented with engaging, hands-on field experiences. Heinrich uses the plight of Florida's turtles as a case study for teaching ecology and conservation.

## SAVING ONE TURTLE AT A TIME

Raising public awareness about diamondback terrapins is a necessary conservation step, especially in coastal communities where this turtle can be found. Many people may be aware of the decline of the turtle soup fad and the lessening of harvesting stress on diamondback terrapin populations. However, there may be a false sense of an assured rebound for the species now that turtle soup is not a common first course on the dinner table. Instead of soup cravings, many of our other activities and pursuits, commercial as well as recreational, are taking their toll on the terrapin.

Every injured female that is rehabilitated will breed and produce hatchlings for many years. Not a single injured female should be denied veterinary care. Justin and Juliet Blass, my nephew and niece, gave just such an injured female a second chance. They had never seen a turtle in the water near their home in Jamesport, New York. They observed a lethargic female terrapin tumbling around in the onshore waves off their beachfront property. The turtle was having difficulty swimming and kept getting swept back to shore. Mystified by seeing a turtle in Peconic Bay and sensing that the turtle was in trouble, the Blass children and their father made an emergency call to the Riverhead Aquarium, where staff guided them to a wildlife rehabilitation center. Packed in a Tupperware container, the female terrapin was brought in for observation. When she was examined, a large gash was found on her plastron, from a suspected boat strike. If this female had been left to her own devices, she probably would not have survived. Transporting terrapins to treatment centers takes time and effort, but informed, caring individuals can make a difference. This injured turtle is on the mend. She was given a chance that she would have been denied if she had not been rescued from the shoreline.

## Founding of the Diamondback Terrapin Working Group

In 1994, thirty-three individuals from thirteen states attended the First Workshop on the Ecology, Status and Management of the Diamondback Terrapin (*Malaclemys terrapin*) at the Savannah River Ecology Laboratory. As compiled by Seigel and Gibbons (1995), the recommendations that resulted from the workshop included funding of research in the following areas:

- Demography.
- Genetic studies.
- Habitat use.
- Movement patterns and home range size.
- Ecology of juveniles.
- Long-term life history studies.
- Taxonomic studies on subspecies.
- Behavioral ecology.

Although there was an overall sense that terrapin numbers were declining, workshop participants agreed that there was a lack of information on the population status of the diamondback terrapin in most states and that there was insufficient evidence to propose listing diamondback terrapins under the Endangered Species Act. The workshop also highlighted the most serious threats to diamondback terrapins that were recognized in 1994; crab pots and habitat loss were at the top of the list. Other pressures on the population were identified: commercial harvesting for food and pet trades, road mortality, boat strikes, and predation by raccoons.

The Second Workshop on the Ecology, Status and Conservation of Diamondback Terrapins was held in 2000 at the Wetlands Institute in Stone Harbor, New Jersey, where Roger Wood served as host. Discussion and analysis of the terrapin's plight continued. Ten years after the first meeting, the third workshop in the series was held in Jacksonville, Florida. Joe Butler and George Heinrich, organizers of the workshop, urged participants to think in terms of wide-range conservation. At the third workshop, a national Diamondback Terrapin Working Group (DTWG) was formed by unanimous vote of attendees. The mission statement and objectives, as drafted by Butler, Heinrich, and Roosenburg and approved by the working group are as follows:

### Mission Statement

We are a group of individuals from academic, scientific, regulatory and private institutions/organizations working to promote the conservation of the diamondback terrapin, the preservation of intact, wild terrapin populations and their associated ecosystems throughout their range. The Diamondback Terrapin Working Group is committed to and supports research, management, conservation and education efforts with the above goals in mind.

## Objectives

1. To advocate and promote sound, scientifically based survey and population studies that can identify demographic trends and identify causal factors contributing to changes in population size, growth and structure.
2. To identify situations which threaten the existence of terrapin populations and take necessary steps to remedy those situations.
3. To maintain a database of the known terrapin populations which are or have been studied with specific attention to changes in population growth rate.
4. To provide advice and recommendations for the research direction and effective management and conservation of terrapins throughout their range.
5. To promote and assist educational programs that focus on terrapin conservation or which use the terrapin as a model organism to promote environmental awareness and stewardship.
6. To meet once every three years as the "Workshop on the Ecology, Status and Conservation of Diamondback Terrapins." The Diamondback Terrapin Working Group will hold an open meeting during the workshop to conduct general business.
7. To serve as a source of information on terrapins and their associated habits. One aspect of this will be to maintain a bibliography of all known scientific publications concerning diamondback terrapins.

## Regional Strategies

Although range-wide planning is critical, regional strategies are equally important. There are many localized threats that affect terrapins in certain areas more than others. The best model I have seen for general recommendations for diamondback terrapin conservation are those of the Maryland Diamondback Terrapin Task Force, established in January 2001 by Parris N. Glendenning, governor of Maryland at the time. Some recommendations can be used in any region, but some are unique to Chesapeake Bay. After careful consideration of the data at hand and the subsequent conclusion that this historically notable species was in decline, the task force drafted a report in September 2001 with interim recommendations that are posted at the web site of the Terrapin Institute (http://www.terrapininstitute.org/TFR2001.htm). This report provides a comprehensive listing of proposed actions to augment terrapin conservation efforts in Chesapeake Bay. Key recommendations include:

- Moratorium on commercial harvest.
- Stock assessment.
- Establishing and enforcing legal size limits.
- Further limits to the time for harvest to coincide with the historic ban (April 1–November 1).
- Limitation in the issue of new commercial harvest licenses.
- Banning of commercial crab pots in tributaries.
- Development of cost-effective BRDs.
- Relisting of terrapins as "In Need of Conservation."
- Identification and protection of nesting beaches.
- Enactment of legislation and administrative policy for shoreline erosion control and beach strand preservation.
- Establishment of management policy for shoreline stabilization that will create, rather than destroy, terrapin nesting habitat.
- Continue and expand the program for head starting/repatriation for public education.
- Establish an annual Diamondback Terrapin Day in May to promote public awareness and appreciation of terrapins and stewardship of Chesapeake Bay.
- Enact humane treatment regulations for terrapin handling and shipment.

As we consider recommendations for diamondback terrapin conservation, we might also learn some lessons from research and management of their marine cousins. For sea turtles, high survival rates are needed at all life stages to ensure that individuals will reach reproductive age and maintain populations. In the case of the diamondback terrapin, we must understand habitat requirements for all life stages, and these habitats must be protected. There should be at least a temporary halt to commercial harvest until populations can be adequately characterized. Similar to the conditions that led to the crash of sea turtle populations, sustainable harvest of long-lived, late-maturing turtle species may be impossible.

In some cases, we may have to reevaluate guidelines for recreation. It may be difficult to mandate and enforce seasonal closure of terrapin nesting sites to beachgoers or to limit recreational crabbing but there are precedents in which similar strategies are used to manage other species. Anglers and visitors to Cape Cod have come to expect seasonal beach closures to ORV traffic when piping plovers are nesting.

The complexity of turtle conservation has been articulated by many researchers (Klemens, 2000). The conservation issues that affect diamondback terrapins also impact many other turtle species. Highly manipulative measures that are being contemplated or designed to increase the local terrapin population, such as headstarting, relocation, or repatriation are controversial and have had only limited success with other species. Furthermore, these will not be viable conservation strategies if suitable habitat is missing or destroyed.

High hatchling and juvenile mortality means that adults must live long lives and continue reproducing in order to stabilize the population. However, many threats specifically affect the reproductive-aged adults. We must pay attention to diamondback terrapin life history traits and design conservation strategies accordingly. We deal with a species that exhibits delayed maturity, longevity, site fidelity, use of different habitats during different life stages, low nesting success, and increased or high levels of depredation. We cannot expect diamondbacks to respond rapidly to multiple anthropogenic stresses. A multifaceted conservation approach will be necessary; long-term monitoring will be a key component in assessing the results of any program.

Some conservation efforts appear to be paying off. In some areas, the local population of diamondback terrapins has noticeably increased since the dangerously low levels of the early and mid 1900s and currently appears stable. But even with the protective measures that are already in place, many other factors are having a profound impact on potential rebound of the terrapin, and vigilance will be necessary in order to assure successful recovery of these turtles. Although we may not have all the demographic data that are necessary to produce conservation and management guidelines and legislate optimal regulations, let us not suffer from a paralysis of analysis. We can and should move forward with some immediate measures, such as requirements for TEDs/BRDs. Some researchers have also suggested a requirement for biodegradable panels on all crab pots to further reduce bycatch in ghost pots. Protecting natural nests from predators by using nest protectors/predator excluders can also be implemented without a great deal of expense.

The diamondback terrapin's ancestors survived a mass extinction 65 million years ago that completely demolished other reptiles such as the dinosaur. Each year, many terrapins survive six months or more of suspended life, dormant under winter ice and freezing mud. The question remains: Can terrapins survive our interference in their daily and seasonal activities?

## *Terrapins for the Future*

The diamondback terrapin is known outside the world of culinary arts, biology, herpetology, and fishing. Terrapins have also been used in the marketing arena; "terrapin" is in the name or logo for many different commercial enterprises and products such as software companies, puppet theaters, greeting-card companies, home alarm companies, restaurants, jewelry artisans, tile manufacturers, skin and body products, dog kennels and breeders, wet-suit brands, tree nurseries, music companies, and others. Whether these names and logos refer to the diamondback terrapin in particular or to the generic use of the term "terrapin" to describe a turtle is often not clear.

Turtles, in general, are reptiles to which people can readily relate. They are revered in many cultures and are the subject of a diverse array of stories and

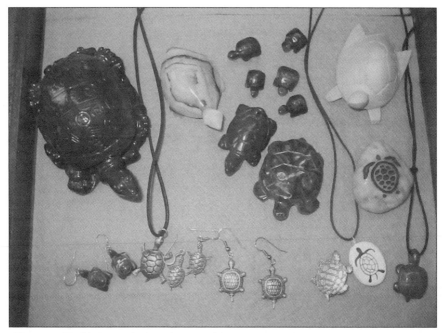

*Fig. 6.6. Turtle/terrapin jewelry, carvings, and statues.*

legends. Children are exposed to them in a friendly and engaging manner in picturebooks, as plastic or stuffed toys, and in the form of television characters. With the exception of the snapping turtle, *Chelydra serpentina*, which has an undeserved reputation as a mean-tempered and dangerous species, turtles are generally considered to be harmless, rather helpless creatures. Although not cuddly or warm, they are nevertheless liked by most people. They tend to be shy, mind their own business, and are not loud or disruptive. They have pleasing body symmetry and charismatic facial features. Artists have rendered many different species of turtle, including the terrapin, in a variety of media. Earrings and jewelry are available from many sources. Terrapins and other types of turtles have been sculpted from myriad materials (fig. 6.6). It is a rare household that does not have a "turtle" of some kind residing within.

Let us hope that our aesthetic appreciation of these creatures will extend beyond the incorporation of terrapin images and the use of terrapin logos. Our relationship with this turtle must motivate us to initiate further research, conservation, and education efforts to ensure that we leave more than just the rendering or representation of the terrapin for future generations.

# Bibliography

Alderton, D. 1988. *Turtles & Tortoises of the World.* New York, N.Y.: Facts on File Publications.

Allen, E. R., and W. T. Neill. 1952. The diamondback terrapin. *Florida Wildlife* 6(6):8,42.

Allen, F. J., and R. A. Littleford. 1955. Observations on the feeding habits and growth of immature diamondback terrapins. *Herpetologica* 11:77–80.

Amato, G. D., J. L. Behler, B. W. Tryon, and D. W. Herman. 1997. Molecular variation in the bog turtle, *Clemmys muhlenbergii*. In *Conservation, Restoration and Management of Tortoises and Turtles—An International Conference*, State University of New York, Purchase, N.Y., pp. 259-22. New York, N.Y.: New York Turtle and Tortoise Society.

Argyriou, A., M. Krause, and R. L. Burke. 2004. Multiple paternity in diamondback terrapins at Jamaica Bay Wildlife Refuge. Presented at the Northeast Natural History Conference VIII, New York State Museum, Albany, N.Y.

Arndt, R. G. 1991. Predation on hatchling diamondback terrapin, *Malaclemys terrapin* (Schoepff), by the ghost crab, *Ocypode quadrata* (Fabricius). *Florida Scientist* 54(3/4):215–217.

Arndt, R. G. 1994. Predation on hatchling diamondback terrapin, *Malaclemys Terrapin* (Schoepff), by the ghost crab *Ocypode Quadrata* (Fabricius). II. *Florida Scientist* 57(1):1–5.

Auger, P. J., and P. Giovannone. 1979. On the fringe of existence: Diamondback terrapins at Sandy Neck. *Cape Naturalist* 8(3).

Avise, J. C., B. W. Bowen, T. Lamb, A. B. Meylan, and E. Bermingham. 1992. Mitochondrial DNA evolution at a turtle's pace: Evidence for low genetic variability and reduced microevolutionary rate in the Testudines. *Mol. Biol. Evol.* 9(3):457–473.

Babcock, H. L., 1926. The diamondback terrapin in Massachusetts. *Copeia* 150:101–104.

Babcock, H. L., 1938. *Field Guide to New England Turtles.* Natural History Guides, No. 2. Boston: Boston Society of Natural History.

Baker, P. J., J. P. Costanzo, R. Herlands, M. Draud, R. C. Wood, and J. R. E. Lee. 2004. Cold-hardiness of terrestrially hibernating hatchlings of the northern diamondback terrapin, *Malaclemys terrapin terrapin*. Presented at Third Workshop on the Ecology, Status and Conservation of Diamondback Terrapins, Jacksonville, Fla.

Bauer, A. M., and R. A. Sadlier. 1992. A device for separating fecal samples of a mollusc-feeding turtle, *Malaclemys terrapin*. *Herpetol. Rev.* 23(4):113–115.

Bauer, B. A., and M. J. Draud. 2004. Effect of syzygy on nesting biorhythms of an estuarine turtle, *Malaclemys terrapin*. Presented at The Northeast Natural History Conference VIII, New York Museum. Albany, N.Y.

Bels, V. L., J. Davenport, and S. Renous. 1995. Drinking and water expulsion in the diamondback turtle *Malaclemys terrapin*. *J. Zool., Lond.* 236:483–497.

Bels, V. L., J. Davenport, and S. Renous. 1998. Food ingestion in the estuarine turtle *Malaclemys terrapin*: Comparison with the marine leatherback turtle *Dermochelys coriacea*. *J. Mar. Biol. Assoc.* 78:953–972.

Berrill, M., and D. Berrill. 1981. *A Sierra Club Naturalist's Guide to the North Atlantic Coast: Cape Cod to Newfoundland*. San Francisco, CA: Sierra Club Books.

Bishop, J. M. 1983. Incidental capture of diamondback terrapin by crab pots. *Estuaries* 6:426–430.

Breininger, D. R., M. J. Barkaszi, D. M. Oddy, and J. A. Provancha. 1998. Prioritizing wildlife taxa for biological diversity conservation at the local scale. *Environ. Manage.* 22:315–321.

Brennessel, B., J. Chadwick, C. Stewart-Swift, and N. Warren. 2004. The importance of salt marsh as a nursery for diamondback terrapins. Presented at the Northeast Natural History Conference VIII, New York State Museum, Albany, N.Y.

Burger, J. 1976. Temperature relationships in nests of the northern diamondback terrapin, *Malaclemys terrapin terrapin*. *Herpetologica* 32(4):412–418.

Burger, J. 1977. Determinants of hatching success in diamondback terrapin, *Malaclemys terrapin*. *American Midland Naturalist* 97(2):444–464.

Burger, J. 2002. Metals in tissues of diamondback terrapin from New Jersey. *Environ. Monit.Assess.* 77:255–263.

Burger, J., and S. D. Garber. 1995. Risk assessment, life history strategies and turtles. *J. Toxicol. Environ. Health* 46(4):483–500.

Burger, J., and W. A. Montevecchi. 1975. Tidal synchronization and nest sight selection in the northern diamondback terrapin *Malaclemys terrapin terrapin* Schoepff. *Copeia* 1:113–119.

Burns, T. A., and K. L. Williams. 1972. Notes on reproductive habits of *Malaclemys terrapin pileata*. *Herpetology* 6(3–4):237–238.

Butler, J. A. 2000. Status and distribution of the Carolina diamondback terrapin, *Malaclemys terrapin centrata*, in Duval County. Florida Fish and Wildlife Conservation Commission. Tallahassee, Fla.

Butler, J. A., C. Broadhurst, M. Green, and Z. Mullin. 2004. Nesting, nest predation and hatchling emergence of the Carolina diamondback terrapin, *Malaclemys terrapin centrata*, in northeastern Florida. *Am. Midland Nat.* 152:145–155.

Butler, J. A., and G. L. Heinrich. 2004. Effectiveness of a bycatch reduction device on crab pots in Florida—Preliminary results. Presented at the Third Workshop on the Ecology, Status and Conservation of Diamondback Terrapins, Jacksonville, Fla.

Cagle, F. R. 1939. A system for marking turtles for future identification. *Copeia* 2:170–173.

Cagle, F. R. 1952. A Louisiana terrapin population (*Malaclemys*). *Copeia* 2:74–76.

Cannon, P., and P. Brooks. 1968. *The Presidents' Cookbook—Practical Recipes from George Washington to the Present*. New York: Funk & Wagnalls.

Carr, A. F., Jr. 1940. *A Contribution to the Herpetology of Florida*. vol. III. Gainseville, Fla.: University of Florida Press.

Carr, A. F., Jr. 1946. Status of the mangrove terrapin. *Copeia* 3(3):170–172.

Carr, A. 1952. *Handbook of Turtles. The Turtles of the United States, Canada, and Baja California*. Ithaca, N.Y.: Cornell University Press.

Charest, B. 1994. Road kill survey. *Resource Manage. Notes: DNR Newslet. Nat. Resource management.* 6(2):4.

Coker, R. E. 1906. The natural history and cultivation of the diamondback terrapin. *North Carolina Geol. Survey* 14:3–69.

Coker, R. E. 1920. The diamond-back terrapin: Past, present and future. *Sci. Month.* 11:171–186.

Congdon, J. D., S. W. Gotte, and R. W. McDiarmid. 1992. Ontogenetic changes in habitat use by juvenile turtles *Chelydra serpentina* and *Chrysemys picta. Can. Field Nat.* 106:241–248.

Cook, R. 1989. A natural history of the diamondback terrapin. *Underwater Nat.* 18(1):25–31.

Cowan, F. B. M. 1971. The ultrastructure of the lachrymal "salt" gland and the Harderian gland in the euryhaline *Malaclemys* and some closely related stenohaline emydines. *Can. J. Zool.* 49:691–697.

Cowan, F. B. M. 1973. Observations on extrarenal excretion by orbital glands and osmoregulation in *Malaclemys terrapin. Comp. Biochem. Physiol.* 48A:489–500.

Davenport, J., and E. A. Macedo. 1990. Behavioural osmotic control in the euryhaline diamondback terrapin *Malaclemys terrapin*: Responses to low salinity and rainfall. *J. Zool. Lond.* 220:487–496.

Davenport, J., and S. H. Magill. 1996. Thermoregulation or osmotic control? Some preliminary observations on the function of emersion in the diamondback terrapin *Malaclemys terrapin* (Latreille). *Herpetol. J.* 6:26–29.

Davenport, J., M. Spikes, S. M. Thornton, and B. O. Kelly. 1992. Crab-eating in the diamondback terrapin *Malaclemys terrapin*: Dealing with dangerous prey. *J. Mar. Biol. Assoc. U.K.* 72:835–848.

Davenport, J., and J. F. Ward. 1993. The effects of salinity and temperature on appetite in the diamondback terrapin *Malaclemys terrapin* (Latreille). *Herpetol. J.* 3(3):95–98.

Dobie, J. L. 1981. The taxonomic relationship between *Malaclemys Gray 1841* and *Graptemys Agassiz, 1857. Tulane Stud. Zool. Bot.* 23:85–102.

Dobie, J. L., and D. R. Jackson. 1979. First fossil record of the diamondback terrapin *Malaclemys terrapin (Emydidae)* and comments on the fossil record of *Chrysemys nelsoni (Emydidae). Herpetologica* 35(2):139–145.

Doody, J. S., P. West, and A. Georges. 2003. Beach selection in nesting pig-nosed turtles, *Carettochelys insculpta. J. Herpetol.* 37(1):178–182.

Douglas, M. S. 1947. *The Everglades: River of Grass.* Sarasota, Fla.: Pineapple Press.

Draud, M., S. Zimnavoda, T. King, and M. Bossert. 2004. Aspects of behavior and ecology of hatchling and juvenile diamondback terrapins. Presented at the Third Workshop on the Ecology, Status and Conservation of Diamondback Terrapins, Jacksonville, Fla.

Draud, M. J., and M. Bossert. 2002. Semi-aquatic habitat use by diamondback terrapin hatchlings in a New York estuary. Presented at the Northeast Natural History Conference VII, New York State Museum, Albany, N.Y.

Draud, M. J., M. Bossert, and S. Zimnavoda. 2004. Predation on hatchling and juvenile diamondback terrapins (*Malaclemys terrapin*) by the Norway rat (*Rattus norvegicus*). *Herpetology* 38(3):467–470.

Dunson, M. K., and W. A. Dunson. 1975. The relation between plasma Na concentration and salt gland Na-K ATPase content in the diamondback terrapin and the yellow-bellied sea snake. *J. Comp. Physiol.* 101:89–97.

Dunson, W. A. 1970. Some aspects of electrolyte and water balance in the estuarine reptiles, the diamondback terrapin, American and "salt water" crocodiles. *Comp. Biochem. Physiol.* 32:161–174.

Dunson, W. A. 1976. Salt glands in reptiles. *Biology of the Reptilia*, ed., C. Gans, 413–445. New York: Academic Press.

Dunson, W. A. 1985. Effect of water salinity and food salt content on growth and sodium efflux of hatchling diamondback terrapins (*Malaclemys*). *Physiol. Zool.* 58:736–747.

Dunson, W. A., and F. J. Mazzotti. 1989. Salinity as a limiting factor in the distribution of reptiles in Florida Bay: A theory for the estuarine origin of marine snakes and turtles. *Bull. Mar. Sci.* 44(1):229–244.

Ernst, C. H., and B. R. Bury. 1982. *Malaclemys, M. terrapin. Catalogue Am. Amphib. Reptiles* 299:1–4.

Ernst, C. H., J. H. Lovich, and R. W. Barbour. 1994. *Turtles of the United States and Canada.* Washington, D.C.: Smithsonian Institution Press.

Estep, B. 2004. Evaluation of the seasonal movements and habitat use of a South Carolina diamondback terrapin population using sonic tracking. Presented at the Third Workshop on the Ecology, Status and Conservation of Diamondback Terrapins, Jacksonville, Fla.

Ewert, M. A., and C. E. Nelson. 1991. Sex determination in turtles: Diverse patterns and some possible adaptive values. *Copeia* 1:50–69.

Feinberg, J. A., and R. L. Burke. 2003. Nesting ecology and predation of diamondback terrapins, *Malaclemys terrapin*, at Gateway National Recreation Area, New York. *J. Herpetol.* 37(3):517–526.

Ferri, V. 2002. *Turtles and Tortoises: A Firefly Guide.* Richmond Hill, Ontario: Firefly Books, Ltd.

Finneran, L. C. 1948. Diamond-back terrapin in Connecticut. *Copeia* 2:138.

Garber, S. D. 1986. Diamondback terrapin exploitation. *Plastron Papers* XVII(6):18–22.

Garber, S. D., and J. Burger. 1995. A 20-yr study documenting the relationship between turtle decline and human recreation. *Ecological Applications* 5(4):1151–1162.

Gauthier, D., D. Gregson, D. Wiktor, R. Chambers, and M. Hill. 2000. Terrapin research in southern New England. Presented at the Second Workshop on the Ecology, Status and Conservation of Diamondback Terrapins, The Wetlands Institute, Stone Harbor, N.J.

Giambanco, M. R. 2002. Comparison of viability rates, hatchling survivorship, and sex ratios of laboratory-and field-incubated nests of the estuarine, emydid turtle *Malaclemys terrapin*, Master's thesis, Biology Department, Hofstra University.

Gibbons, J. W. 1987. Why do turtles live so long? *BioScience* 37(4):262–269.

Gibbons, J. W., M. E. Dorcas, J. D. Willson, and J. L. Greene. 2004. Lessons from two decades of tracking terrapins in tidal creeks: Kiawah Island, South Carolina. Presented at the Third Workshop on the Ecology, Status and Conservation of Diamondback Terrapins, Jacksonville, Fla.

Gibbons, J. W., J. E. Lovich, A. D. Tucker, N. N. FitzSimmons, and J. L. Greene. 2001. Demographic and ecological factors affecting conservation and management of the diamondback terrapin (*Malaclemys terrapin*) in South Carolina. *Chelonian Conserv. Biol.* 4(1):66–74.

Gibbs, J. P., and D. A. Steen. 2005. Trends in sex ratios of turtles in the United States: Implications of road mortality. *Conserv. Biol.* 19(2):552–556.

Gilles-Baillien, M. 1970. Urea and osmoregulation in the diamondback terrapin *Malaclemys centrata* (Latreille). *J. Exp. Biol.* 52:691–697.

Gilles-Baillien, M. 1973a. Osmotic regulation in various tissues of the diamondback terrapin *Malaclemys centrata centrata* (Latreille). *J. Exp. Biol.* 59:39–43.

Gilles-Baillien, M. 1973b. Seasonal variations and osmoregulation in the red blood cells of the diamondback terrapin *Malaclemys centrata centrata* (Latreille). *Comp. Biochem. Physiol. A Comp. Physiol.* 46A:505–512.

Gilles-Baillien, M. 1973c. Hibernation and osmoregulation in the diamondback terrapin *Malaclemys centrata centrata* (Latreille). *J. Exp.Biol.* 59:45–51.

Goodwin, C. C. 1994. Aspects of nesting ecology of the diamondback Terrapin (*Malaclemys Terrapin*) Rhode Island, Master's Thesis, zoology, University of Rhode Island.

Guillory, V., and P. Prejean. 1998. Effect of a terrapin excluder device on blue crab, *Callinectes sapidus*, trap catches. *Mar. Fisheries Rev.* 60(1):38–40.

Harris, J. C. 1930. *Uncle Remus: his songs and his sayings.* New York: D. Appleton.

Hart, K. 2000. Diamondbacks in North Carolina: Insights from terrapin population modeling, radio telemetry, and mark–recapture efforts. Presented at the Second Workshop on the Ecology, Status and Conservation of Diamondback Terrapins, The Wetlands Institute, Stone Harbor, N.J.

Hart, K. M. 1999. Declines in diamondbacks: Terrapin population modeling and implications for management, Master's Thesis, Nicholas School of the Environment, Duke University.

Hart, K. M. 2004. Population biology of the diamondback terrapin (*Malaclemys terrapin*): Integrating ecology and genetics and conserve distinct population segments. Third Workshop on the Ecology, Status and Conservation of Diamondback Terrapins, Jacksonville, Fla.

Hart, K. M., S. S. Heppell, L. B. Crowder, and P. J. Auger. June 2000. Declines in diamondbacks: Terrapin population modeling and implications for management. 80th Annual Meeting American Society of Ichthyologists and Herpetologists, Universidad Autonoma de Baja California Sur, La Paz, B.C.S., Mexico June. Abstract 186.

Harwood, V. J., J. Butler, D. Parrish, and V. Wagner. 1999. Isolation of the fecal coliform bacteria from the diamondback terrapin (*Malaclemys terrapin centrata*). *Applied and Env. Microbiol.* 65:865–867.

Hauswaldt, J. S. 2004. Site fidelity, population genetics and mating pattern in East Coast diamondback terrapins. Presented at the Third Workshop on the Ecology, Status and Conservation of Diamondback Terrapins, Jacksonville, Fla.

Hauswaldt, J. S., and T. C. Glenn. 2003. Microsatellite DNA loci from the Diamondback terrapin (*Malaclemys terrapin*). *Mol. Ecol. Notes* 3:174–176.

Hauswaldt, J. S., and T. C. Glenn. 2005. Population genetics of the diamondback terrapin (*Malaclemys terrapin*). *Mol. Ecol.* 14:723–732.

Hedges, B. S., and L. L. Poling. 1999. A molecular phylogeny of reptiles. *Science* 283:998–1001.

Hildebrand, S. E. 1929. Review of experiments on artificial culture on diamondback terrapin. *Bull. Bureau Fisheries* 45:25–70.

Hildebrand, S. E. 1932. Growth of diamond-back terrapins: Size attained, sex ratio and longevity. *Zoologica* IX(15):551–563.

Hildebrand, S. E., and C. Hatsel. 1926. Diamondback terrapin culture at Beaufort, N. Carolina. *Econ. Circ.* 60:1–20.

Hogan, J. L. 2003. Occurrence of the diamondback terrapin (*Malaclemys terrapin littoralis*) at South Deer Island in Galveston Bay, Texas, April 2001–May 2002. U. S. Geological Survey. Report 03–022, Austin, Tex.

Hoyle, M. E., and J. Whitfield Gibbons. 2000. Use of a marked population of diamondback terrapins (*Malaclemys terrapin*) to determine impacts of recreational crab pots. *Chelonian Conserv. Biol.* 3(4):735–37.

Hunt, C., K. Morgan, and B. Brennessel. 2002. Do northern diamondback terrapins (*Malaclemys terrapin terrapin*) use light onset to set their circaidian activity cycles. Presented at the Eastern New England Biology Conference, Boston College.

Hurd, L. E., G. W. Smedes, and T. A. Dean. 1979. An ecological study of a natural population of diamondback terrapins (*Malaclemys t. terrapin*) in a Delaware salt marsh. *Estuaries* 2(1):28–33.

Jackson, C. G., Jr., and A. Ross. 1971. Molluscan fouling of the ornate diamondback terrapin, *Malaclemys terrapin macrospilota* Hay. *Herpetologica* 27:341–344.

Jackson, C. G., Jr., A. Ross, and G. L. Kennedy. 1973. Epifaunal invertebrates of the ornate diamondback terrapin, *Malaclemys terrapin macrospilota*. *Am. Midland Nat.* 89:495–497.

Jeyasuria, P., W. M. Roosenburg, and A. R. Place. 1994. Role of P-450 aromatase in sex determination of the diamondback terrapin, *Malaclemys terrapin*. *J. Exp. Zool.* 270:95–111.

Jeyasuria, P., and A. R. Place. 1997. Temperature-dependent aromatase expression in developing diamondback terrapin (*Malaclemys Terrapin*) embryos. *J. Steroid Biochem. Mol. Biol.* 61(3–6:415–25.

Junior League of Lafayette. 1967. *Talk About Good!* Lafayette, La.: Junior League of Lafayette.

Johnson, W. R. J. 1952. Range of *Malaclemmys* [sic] *terrapin* rhizophorarum on the west coast of Florida. *Herpetologica* 8:100.

Kannan, K., H. Nakata, R. Stafford, G. R. Masson, S. Tanabe, and J. P. Giesy. 1998. Bioaccumulation and toxic potential of extremely hydrophobic polychlorinated biphenyl congeners in biota collected at a Superfund site contaminated with Aroclor 1268. *Environ. Sci. Technol.* 32:1214–21.

King, T. L., and S. E. Julian. 2004. Conservation of microsatellite DNA flanking sequence across 13 emydid genera assayed with novel bog turtle (*Glyptemys muhlenbergii*) loci. *Conserv. Genet.* 5:719–725.

Klemens, M. W. 1993. *Amphibians and Reptiles of Connecticut and Adjacent Regions.* Bulletin

No. 112. Hartford, Conn.: State Geological and Natural History Survey of Connecticut DEP Maps and Publications Office.

Klemens, M. W. 2000. *Turtle Conservation*. Washington, D.C.: Smithsonian Institution Press.

Lamb, T., and J. C. Avise. 1992. Molecular and population genetic aspects of mitochondrial DNA variability in the diamondback terrapin, *Malaclemys terrapin. J. Hered.* 83:262–69.

Lamb, T,. and M. F. Osentoski. 1997. On the paraphyly of *Malaclemys*: A molecular genetic assessment. *J. Herpetol.* 31(2):258–265.

Lazell, J. D., Jr. 1976. *This Broken Archipelago—Cape Cod and the Islands, Amphibians and Reptiles*. New York: Quadrangle/The New York Times Book Co.

Lazell, J. D., Jr., and P. J. Auger. 1981. Predation on diamondback terrapin (*Malaclemys terrapin*) eggs by dunegrass (*Ammophila breviligulata*). *Copeia* 3:724–726.

Levin, T. 2003. *Liquid Land. A Journey Through the Florida Everglades*. Athens: University of Georgia Press.

Lovich, J. H., and J. W. Gibbons. 1990. Age at maturity influences adult sex ratio in the turtle *Malaclemys terrapin. Oikos* 59:126–134.

Lovich, J. H., A. D. Tucker, D. E. Kling, J. W. Gibbons, and T. D. Zimmerman. 1991. Behaviour of hatchling diamondback terrapins (*Malaclemys terrapin*) released in a South Carolina marsh. *Herpetol. Rev.* 22(3):81–83.

Mealey, B. K., G. M. Parks, J. D. Baldwin, and M. R. J. Forstner. 2004. The insular ecology of the *Malaclemys terrapin macrospilota* and *M. t. rhizophorarum* of Florida Bay and the Florida Keys. Presented at the Third Workshop on the Ecology, Status and Conservation of Diamondback Terrapins, Jacksonville, Fla.

Middaugh, D. 1981. Reproduction ecology and spawning periodicity of the Atlantic silverside *Menidia menidia. Copeia* 4:766–776.

Miller, L. A. 2001. Population status and potential storm dispersal events of *Malaclemys terrapin* in Florida Bay, Florida. Master's thesis. Department of Biology, Florida Atlantic University.

Mitro, M. G. 2003. Demography and viability analyses of a diamondback terrapin population. *Can. J. Zool.* 81:716–726.

Montevecchi, W. A., and J. Burger. 1975. Aspects of the reproductive biology of the Northern diamondback terrapin *Malaclemys terrapin terrapin. Am. Midland Naturalist* 94(1):166–178.

Morris, A. D. 2004. Texas abandoned crab trap removal program, 2002 to 2004. Presented at the Third Workshop on the Ecology, Status and Conservation of Diamondback Terrapins, Jacksonville, Fla.

Muehlbauer, E. 1987. Field and laboratory studies of the cctivity in the turtle *Malaclemys terrapin terrapin*. Master's Thesis, Department of Biology, New York University.

Nelson, D. H., J. J. Dindo, and R. C. Wood. 2000. Preliminary status of diamondback terrapins in the vicinity of Mobile Bay, Alabama. Presented at the Second Workshop on the Ecology, Status and Conservation of Diamondback Terrapins, The Wetlands Institute, Stone Harbor, N. J.

Page, D., and B. Brennessel. 2005. Multiple Paternity in Diamondback Terrapins. Presented at the National Conference on Undergraduate Research, Alexandria, Va.

Palmer, W. M., and C. L. Cordes. 1988. Habitat Suitability Index Models: Diamondback Terrapin (Nesting)—Atlantic Coast. U.S. Department of the Interior, Fish and Wildlife Service, Research Development, National Wetlands Research Center, Washington, D.C. Report 82(10.151).

Pennisi, E. 2004. Neural beginnings for the turtle's shell. *Science* 303:95.

Perry, H. M. 2004. An overview of the regional program to remove derelict and abandoned crab traps from coastal waters of the Gulf of Mexico. Presented at the Third Workshop on the Ecology, Status and Conservation of Diamondback Terrapins, Jacksonville, Fla.

Pieau, C., and M. Dorizzi. 2004. Oestrogens and temperature-dependent sex determination in reptiles: All is in the gonads. *J. Endocrinol.* 181:367–377.

Pitler, R. 1985. *Malaclemys terrapin terrapin* (Northern diamondback terrapin). *Herpetol. Rev.* 16(3):82.

Place, A. R., J. Lang, S. Gavasso, and P. Jeyasuria. 2001. Expression of P450 (arom) in *Malaclemys terrapin* and *Chelydra serpentina*. *J. Exp. Zool.* 290(7):673–690.

Pope, C. H. 1946. *Turtles of the United States & Canada*. New York: Alfred A. Knopf.

Porter, K. R. 1972. Reproductive Adaptations of Reptiles. In *Herpetology*, Philadelphia, Pa.: W. B. Saunders Co.

Rieppel, O., and R. R. Reisz. 1999. The origin and early evolution of turtles. *Ann. Rev. Ecol. Syst.* 30:1–22.

Robinson, G. D., and W. A. Dunson. 1976. Water and sodium balance in the estuarine diamondback terrapin (*Malaclemys*). *J. Comp. Physiol.* 105:129–152.

Rombauer, I. S., M. R. Becker, and E. Becker. 1975. *Joy of Cooking*. Indianapolis, Ind.: Bobbs-Merrill.

Roosenburg, W. M. 1994. Nesting habitat requirements of the diamondback terrapin: A geographic comparison. *Wetland J.* 6(2):8–11.

Roosenburg, W. M. 1996. Maternal condition and nest site choice: An alternative for the maintenance of environmental sex determination? *Am. Zool.* 36(2):157–168.

Roosenburg, W. M., W. Cresko, M. Modesitte, and M. B. Robbins. 1997. Diamondback terrapin (*Malaclemys terrapin*) mortality in crab pots. *Conserv. Biol.* 11(5):1166–1172.

Roosenburg, W. M., and A. E. Dunham. 1997. Allocation of reproductive output: Egg- and clutch-size variation in the diamondback terrapin. *Copeia* 2:290–297.

Roosenburg, W. M., and J. P. Green. 2000. Impact of a bycatch reduction device on diamondback terrapin and blue crab capture in crab pots. *Ecol. Appli.* 10(3):882–89.

Roosenburg, W. M., K. L. Haley, and S. McGuire. 1999. Habitat selection and movements of diamondback terrapins, *Malaclemys terrapin*, in a Maryland estuary. *Chelonian Conserv. Biol.* 3(3):425–429.

Roosenburg, W. M., and K. C. Kelley. 1996. The effect of egg size and incubation temperature on growth in the turtle, *Malaclemys terrapin*. *J. Herpetol.* 30(2):198–204.

Ross, A., and C. G. Jackson, Jr. 1972. Barnacle fouling of the ornate diamondback terrapin, *Malaclemys terrapin macrospilota*. *Crustaceana* 22:203–205.

Rudloe, J. 1979. A tale of two turtles. In *Time of the Turtle*, pp. 92–161. New York: Alfred A. Knopf.

Seigel, R. A. 1980a. Growth rates, sex ratio, and population structure of diamondback terrapins, *Malaclemys terrapin*, from the Atlantic coast of Florida. In *Proceedings of the 28th Annual Meeting of the Herpetologists' League; 23rd Annual Meeting of the Society for the Study of Amphibians and Reptiles*. (abstract) pp. 87–88.

Seigel, R. A. 1980b. Nesting habits of diamondback terrapins (*Malaclemys terrapin*) on the Atlantic Coast of Florida. *Trans. Kansas Acad. Sci.* 83(4):239–246.

Seigel, R. A. 1980c. Courtship and mating behavior of the diamondback terrapin *Malaclemys terrapin tequesta. J. Herpetol.* 14:420–421.

Seigel, R. A. 1980d. Predation by raccoons on diamondback terrapin *Malaclemys terrapin tequesta. J. Herpetol.* 14 87–89.

Seigel, R. A. 1983. Occurrence and effects of barnacle infestations on diamondback terrapins (*Malaclemys terrapin*). *Am. Midland Naturalist* 109:34–39.

Seigel, R. A. 1984. Parameters of two populations of diamondback terrapins (*Malaclemys terrapin*) on the Atlantic Coast of Florida. In *Vertebrate Ecology and Systematics: A Tribute to Henry S. Fitch*, ed. R. A. Seigel, I. E. Hunt, J. L. Knight, L. Malaret, and N. L. Zuschiag, pp. 77–87. Museum of Natural History, Lawrence, Kans.: The University of Kansas.

Seigel, R. A. 1993. Apparent long-term decline in diamondback terrapin population at the Kennedy Space Center, Florida. *Herpetol. Rev.* 24:102–103.

Seigel, R. A., and J. W. Gibbons, J. 1995. Workshop on the ecology, status, and management of the diamondback terrapin (*Malaclemys terrapin*). *Chelonian Conserv. Biol.* 1(3):240–243.

Shine, R. 2004. Seasonal shifts in nest temperature can modify the phenotypes of hatchling lizards, regardless of overall mean incubation temperature. *Funct. Ecol.* 18:43–49.

Spagnoli, J. J., and B. I. Marganoff. 1975. New York's marine turtle. *Conservationist* 29:17–19.

Tucker, A., and N. FitzSimmons. 1992. A device for separating fecal samples of a mollusc-feeding turtle, *Malaclemys terrapin. Herpetol. Rev.* 23(4):113–115.

Tucker, A., J. W. Gibbons, and J. L. Greene. 2001. Estimates of adult survival and migration for diamondback terrapins: Conservation insight from local extirpation within a metapopulation. *Can. J. Zool.* 79:2100–2209.

Tucker, A., R. S. Yeomans, and J. W. Gibbons. 1997. Shell strength of mud snails (*Ilyanassa obsoleta*) may deter foraging by diamondback terrapins (*Malaclemys terrapin*). *Am. Midland Naturalist* 138:224–229.

Tucker, A. D., N. N. FitzSimmons, and W. Gibbons, J. 1995. Resource partitioning by the estuarine turtle *Malaclemys terrapin*: Trophic, spatial, and temporal foraging constraints. *Herpetologica* 51(2):167–181.

Watters, C. F. 2004. A review of rangewide regulations pertaining to diamondback terrapins. Presented at the Third Workshop on the Ecology, Status and Conservation of Diamondback Terrapins, Jacksonville, Fla.

Whitelaw, D. M., and R. N. Zajac. 2002. Assessment of prey availability for diamondback terrapins in a Connecticut salt marsh. *Northeastern Naturalist* 9(4):407–418.

Wood, C. R. 1977. Evolution of the emydine turtles *Graptemys* and *Malaclemys* (Reptilia, Testudines, Emydidae). *Journal of Herpetology* 11(4):415–421.

Wood, R. 1992. Mangrove terrapin. In *Amphibians and Reptiles. Rare and Endangered Biota of Florida*, Ed. P. E. Moler, pp. 204–209. Gainesville, Fla: University Press of Florida.

Wood, C. R. 1997. The impact of commercial crab traps on Northern diamondback terrapins, *Malaclemys terrapin terrapin*. In *Conservation, Restoration and Management of Tortoises and Turtles—An International Conference*, pp. 21–27. New York, N.Y.: State University of New York, New York Turtle and Tortoise Society.

Wood, R. C., and R. Herlands. 1995. Terrapins, tires and traps: Conservation of the Northern diamondback terrapin (*Malaclemys terrapin terrapin*) on the Cape May Peninsula, New Jersey, USA. In *International Congress of Chelonian Conservation*, pp. 254–256. Gonfaron, France.

Yearicks, E. F., C. R. Wood, and W. S. Johnson. 1981. Hibernation of the Northern diamondback terrapin, *Malaclemys terrapin terrapin*. *Estuaries* 4:78–81.

# Index

Page numbers in **bold** indicate figures or tables.

BARBARA BRENNESSEL, Ph.D. is Professor Emerita at Wheaton College where she retired from a teaching and research career in 2013. Although she trained as a biochemist, her experiences as a homeowner on Cape Cod, as well as her membership on the board of several Wellfleet and Cape-wide environmental organizations, led her to pursue research and conservation work outside of the laboratory. She volunteers for agencies and nonprofits that survey horseshoe crabs, count river herring, and rescue stranded marine mammals and cold-stunned sea turtles. She is the author of *Diamonds in the Marsh: A Natural History of the Diamondback Terrapin*; *Good Tidings: The History and Ecology of Shellfish Aquaculture in the Northeast*; *The Alewives' Tale: The Life History and Ecology of River Herring in the Northeast*; and a children's book, *The Adventures of Allie the Alewife*. She is coauthor of *Tidal Water: A History of Wellfleet's Herring River*.